PREFACE

In this book I have attempted to sketch in brief form the molecular mechanisms by which living cells transform energy for their various activities and for their growth and replication. Because exchanges of energy are fundamental to all activities of cells, this book touches on most of the important areas of dynamic biochemistry and may indeed be regarded as an introduction to biochemistry.

I wrote this book for college undergraduates beginning the study of cell biology or molecular biology, to be used as a supplement to textbooks in general biology. However, I have tried to make this new edition rigorous enough in its fundamentals to be useful, at least as a review, for more advanced students of biology, including premedical and medical students. The book presupposes a course in general chemistry, as well as a concurrent biology course.

The book first outlines the basic principles governing energy exchanges in simple enzymatic reactions. These principles are then applied to the ATP-ADP system as the carrier of chemical energy in the cell. The enzymatic reactions which yield phosphate bond energy, such as photosynthesis, fermentation, and respiration, as well as the basic processes requiring phosphate bond energy, biosynthesis, active transport, and contractile processes, are then analyzed. The concept that information is related to energy is developed as a basis for considering the biosynthesis and role of DNA, RNA, and protein in the transmission and expression of genetic information.

This edition of *Bioenergetics* represents a substantial revision and updating, particularly of the chapters on thermodynamics, photosynthesis, active transport, and the biosynthesis of nucleic acids and proteins.

My sincere thanks go to the many students and teachers who have written helpful letters since the first edition appeared. As before, I will continue to welcome comments and criticisms from all.

Sparks, Maryland A. L. L.
June 1971

CONTENTS

1

THE FLOW OF ENERGY IN THE BIOLOGICAL WORLD

Energy is most simply defined as the capacity to do work. There are different kinds or states of energy: potential, kinetic, thermal, electrical, and radiant energy, among others. There are also different kinds of work, such as mechanical, electrical, and osmotic work. The transformation of energy from one type to another and the efficiency of the conversion of energy into work are of central importance in the study of physics and chemistry. When the physical scientist studies a physical or chemical change the first questions he asks are: What forces caused the change to occur? Why did the change come to a stop? Could the occurrence and nature of the change have been predicted? These are very basic questions because every physical or chemical process is the result of the action of an unbalanced force, and a force, in turn, is the product or result of the movement of energy. From these considerations we see that all physical and chemical processes are ultimately the result of the application, movement, or transformation of energy. That area of physical science which deals with the exchanges of energy in collections of matter is known as *thermodynamics*.

Today, all scientists agree that the laws of physics and chemistry, including the principles of thermodynamics, also hold in the biological world: there can be no vitalism or black magic by which living organisms sustain and perpetuate themselves. Just as thermodynamics is the first and most basic way of analyzing processes involving inanimate matter, so is it fundamental in analyzing the behavior of living organisms. *Bioenergetics* is the term we use to designate the study of energy transformations in living organisms. As an introduction to bioenergetics, let us now survey, in this first chapter, the major processes in which energy is transformed in the world of living organisms. First we shall sketch the successive stages in the flow of energy through the biological macrocosm, and contrast the magnitude of the flow of biological energy on the face of the earth with the energy flow taking place in man-made machines. Then we shall consider energy flow in the biological microcosm, the individual cell. In succeeding chapters we shall examine each aspect of biological energy transformations in much more detail.

1–1 FOOD WEBS AND ENERGY FLOW

If we should trace the ultimate source of the energy utilized by any given organism in its natural habitat, let us say a fox in a given domain of woodland, we would find that there is a hierarchy of organisms, called a *food chain*, that provides the fox with the energy and materials required to sustain his life. This food chain might begin with the photosynthetic cells of green plants, which convert carbon dioxide into new cell material. The plant may in turn be consumed by insect larvae, which may in turn be consumed by toads or birds, which may in turn be consumed by the fox. In any given ecological community of living organisms, many individual food chains are interlocked into a *food web*.

Such a food web consists of several layers of organisms. First we have the *producers*, those cells that can utilize the simplest forms of carbon from the surroundings, such as carbon dioxide; then we have a primary layer of *consumers*, which feed on the producers, followed by further layers of consumers. Finally, to complete the cycle, there are the *decomposers*, the bacteria and fungi, which cause decomposition and decay of dead consumers and thus return simple forms of carbon to the soil and atmosphere.

Living organisms can be classified into two great groups depending on their position in the food web: *autotrophic* and *heterotrophic*. Autotrophic organisms (the name means "self-feeding") are those that can use simple forms of carbon, such as carbon dioxide, from which to build all their cell components. The "producers" at the bottom of the food web are autotrophs; they include many bacteria and algae, as well as higher plants. Heterotrophic organisms ("feeding on others"), on the other hand, cannot utilize simple forms of carbon, such as carbon dioxide, and require more complex forms—organic molecules such as glucose. The consumers and decomposers in the food web are heterotrophs, and they ultimately depend on autotrophs to generate the complex nutrients they require.

Now let us examine this food web and trace the sources of energy for each layer of organism. We will find that the great majority of the producers or autotrophic organisms obtain their energy from sunlight, which they use to convert carbon dioxide into more complex cell materials in the process of *photosynthesis*. Thus most of the producers are *photosynthetic autotrophs*. But when we examine the successive layers of consumers we find that none of them have the capacity to use light energy. Rather most of them obtain the energy they need by the combustion of complex organic molecules, such as glucose, obtained from the producers they consume. In this process, which requires oxygen and is called *respiration*, the simple, small carbon dioxide molecule is the end product. Heterotrophic cells, therefore, obtain energy by degrading complex nutrient molecules to simpler forms.

At each level in the food web, energy is expended to perform various kinds of biological work, such as synthesis of new cell material from simple precursors, movement of materials against gradients, and the work of contraction or motion. However, at each level in the food web we find that there are "friction" losses, so that each time some chemical or physical process occurs there is incomplete conversion of one kind of energy into another. As a result, some of the energy made available to each layer of organisms becomes dissipated in the environment and thus unavailable to do work. Only a small fraction of the solar energy absorbed by the producers in the bottom layer of the food web ever reaches the top layer of ultimate consumers. As these ultimate consumers die, and their tissues are degraded to simple organic products by the decomposers, energy is again lost and dissipated in the environment. Ultimately, then, the flow of energy that begins from the sun and courses through the biological food web finally becomes randomized in the environment (Fig. 1–1).

Now let us examine the three major steps in the flow of biological energy: (1) photosynthesis, (2) respiration, and (3) the performance of biological work.

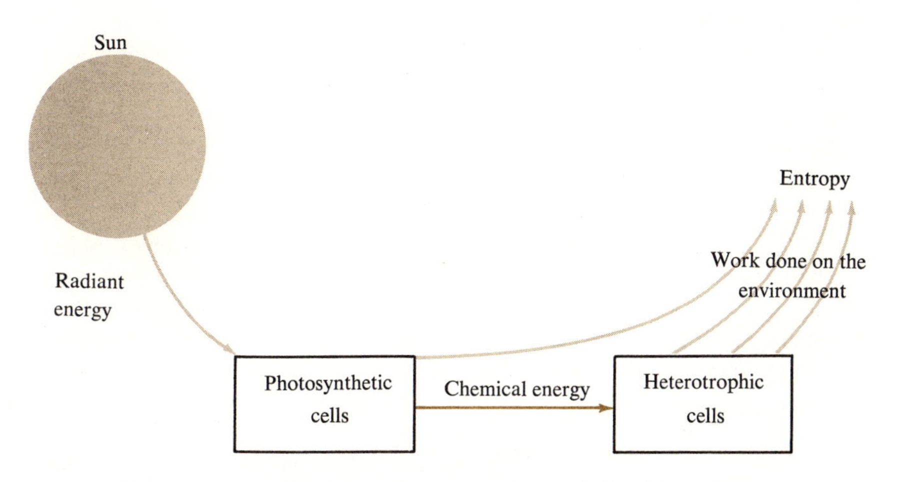

Figure 1–1. The flow of energy through the biosphere.

1–2 SOLAR ENERGY AND PHOTOSYNTHESIS

Visible sunlight, the source of all biological energy, is a form of electromagnetic or radiant energy, which ultimately arises from nuclear energy. In the immensely high temperature of the sun, which is believed to be several million degrees centigrade, a part of the enormous energy locked in the nucleus of hydrogen atoms is released as they are converted into helium atoms (He) and positrons (${}_1e^0$) by thermonuclear fusion.

$$4\,{}^1H \rightarrow {}^4He + 2\,{}_1e^0 + h\nu$$

In this process a quantum of energy in the form of gamma radiation is released. The quantum is represented by the term $h\nu$, in which h is Planck's constant and ν is the frequency of the gamma radiation. After a complex series of reactions in which the gamma radiation is absorbed by the positrons much of the energy of the gamma radiation is emitted in the form of photons, or quanta of light energy. Ultimately then, nuclear fusion reactions in the sun are the source of all biological energy on the earth.

The term "photosynthesis" we tend to associate with the visible world of higher plants: grass, crop plants, and trees. But those macroscopic photosynthetic organisms actually make up but a small fraction of all the known organisms capable of photosynthesis. It has been estimated that some 90 per cent of all the photosynthesis on the earth is carried out in the seas by various kinds of microorganisms, including bacteria, algae, diatoms, and dinoflagellates.

There is another common misconception about photosynthesis in higher plants. Not all cells of a higher plant are capable of photosynthesis. The cells in the roots, stems, and fruits of plants are incapable of photosynthesis; they are heterotrophic and thus resemble animal cells. Only cells with the green pigment chlorophyll are capable of photosynthesis. Moreover, in the dark, when solar energy is unavailable, even these cells function like heterotrophic cells; they must oxidize some of their glucose, at the expense of oxygen, in order to obtain energy in the dark.

Photosynthesis consists of the absorption of radiant energy by chlorophyll and other pigments, the conversion of the absorbed light energy into chemical energy, and the utilization of this chemical energy for reduction of carbon dioxide obtained from the atmosphere to form glucose. In most photosynthetic organisms, particularly the higher plants, molecular oxygen is the other major end product, but in others, such as the photosynthetic bacteria, oxygen is not formed. The overall equation for the photosynthetic formation of glucose and oxygen from carbon dioxide and water in higher plants is most simply given as

$$6CO_2 + 6H_2O \rightarrow C_6H_{12}O_6 + 6O_2 \qquad \Delta G^{0\prime} = +686 \text{ kcal}$$

where the symbol $\Delta G^{0\prime}$ denotes the minimum amount of useful energy that must be furnished by absorbed sunlight to bring about the formation of 1 mole of glucose from one each of CO_2 and H_2O under standard conditions we shall describe later. In chemical thermodynamics the basic unit of energy is the *small* or *gram calorie*, the amount of energy required, in the form of heat, to raise the temperature of 1 g of water at 15.0°C by exactly 1.00°C. The *large calorie* or *kilocalorie* is equal to 1000 small calories.

The large amount of energy that is required to make photosynthesis occur is supplied by the light energy that is captured by the chlorophyll of the leaves. The photosynthetic equation may thus be rewritten to indicate that light quanta are the energy source, as follows.

$$6CO_2 + 6H_2O + n\,h\nu \rightarrow C_6H_{12}O_6 + 6O_2$$

This equation gives us only an overall statement of the photosynthetic process; it says nothing of the mechanism or pathway by which it occurs. Actually photosynthesis in plant cells is a far more complex process than this simple-looking equation might suggest. Although the complete molecular mechanism of photosynthesis is not yet known, there are probably more than a hundred sequential chemical steps in the photosynthetic production of glucose from carbon dioxide and water, each catalyzed by a specific kind of enzyme molecule.

Glucose is not the only product of photosynthesis. Other carbon-containing components of plant cells, such as cellulose, proteins, and lipids, are also produced during photosynthesis. All these substances, which are rich in chemical energy, are ultimately utilized as energy sources by heterotrophic organisms, that is, the consumers that feed on green plants.

1–3 RESPIRATION IN HETEROTROPHIC CELLS

The next stage in the flow of biological energy is the utilization of the energy of carbohydrate, fat, and protein produced in photosynthesis by the heterotrophs, which oxidize these materials with oxygen. Actually heterotrophs require the complex products of photosynthesis for two reasons. First, they need the chemical energy they can obtain by degrading the complex, energy-rich structures of such molecules as glucose. But heterotrophs also need complex carbon compounds, such as glucose, as building blocks for the synthesis of their own cellular components, since they are unable to use carbon dioxide for this purpose.

Heterotrophs include all the organisms of the animal kingdom, many bacteria and fungi, as well as many cells of the plant kingdom. But now we must put aside another common misconception. Although we might think that the familiar large animals of the macroscopic biological world are the predominant heterotrophs in the biosphere, nothing could be further from the truth. It is estimated that over 90 per cent of all the oxygen consumed by all heterotrophs is used by invisible microorganisms of the soil and seas.

Most heterotrophic cells use oxygen they take from the atmosphere to oxidize glucose and other nutrients to form the stable end products carbon dioxide and water. However, some heterotrophs are unable to use oxygen; they degrade glucose to simpler compounds, such as lactic acid, in the absence of oxygen, in the process called *fermentation.* Fermentation products such as lactic acid are ultimately oxidized to CO_2 and H_2O by still other heterotrophic organisms, particularly those that use oxygen. Ultimately, therefore, the cells of the heterotrophic world bring about the complete oxidation of organic nutrients produced by autotrophs, to the end product carbon dioxide. The total process by which foodstuff molecules are ultimately oxidized by heterotrophic cells at the expense of oxygen is referred to as *respiration.*

The chemical equation for oxidation of glucose during respiration is

$$\text{glucose} + 6O_2 \rightarrow 6CO_2 + 6H_2O \qquad \Delta G^{0\prime} = -686 \text{ kcal}$$

We see immediately that this equation is the reverse of that for photosynthesis. Moreover, we note that the complete combustion of 1 mole of glucose can yield a maximum of 686 kcal of useful chemical energy. This is not necessarily the actual yield of work realized by the heterotrophic cell; it is only the theoretical maximum that can be obtained if we have a completely efficient, frictionless machine to harness the energy yielded from the combustion of a mole of glucose.

Although the chemical equation of respiration looks simple, it does not tell us anything of the mechanism or pathway of respiration in heterotrophic cells. Actually there are more than seventy sequential chemical reactions in the oxidation of glucose in heterotrophic cells.

1–4 BIOLOGICAL WORK

We come now to the last great stage in the flow of biological energy, the utilization of chemical energy to do different kinds of cellular work. All living organisms must do work of one kind or another merely to stay alive in an environment that is essentially hostile to them. Some organisms, such as higher vertebrates, may do work on their environment to make it less hostile, whereas others, such as bacteria, overcome the effects of a hostile environment by multiplying rapidly. There are basically three types of work done by living organisms: chemical work, concentration work, and mechanical work.

Chemical work is done by all cells, not only during active growth but also to maintain themselves. The macromolecular components of cells, such as the proteins, nucleic acids, lipids, and polysaccharides, are continuously synthesized from small building-block molecules by the action of enzymes. These processes are collectively called *biosynthesis.* Biosynthesis occurs not only during growth of an organism, when there is a net formation of new cellular material, but also in nongrowing, mature organisms. In the latter, the carbohydrates, proteins, and lipids are constantly being synthesized and degraded in such a way that the rate of formation of the new molecules is exactly balanced by the rate of degradation of the old. Most of the molecular components of living cells exist in such a dynamic steady state.

Whenever we put together a large, ordered structure from simple, randomly disposed units, whether it is a macromolecule or a brick wall, energy is required. To construct a protein molecule, hundreds of individual amino acid molecules must be assembled in the correct sequence and joined in peptide linkage by the action of specific enzymes. To construct a polysaccharide molecule, such as cellulose or starch, hundreds of glucose molecules must be joined in glycosidic linkage. The overall equation for biosynthesis of such macromolecules may be written in generalized form as

$$n(\text{building blocks}) \rightarrow \text{macromolecule} + nH_2O$$

Such biosynthetic reactions, which proceed with loss of water as the building blocks come together, are highly endergonic in the aqueous medium of the cell; they are "uphill" reactions. In Table 1–1 are shown the amounts of useful or free energy required in the biosynthesis of the major types of bonds linking the building blocks of various cell macromolecules.

The second type of cellular work is that required to transport and concentrate substances; it is often, but less accurately, called osmotic work. This kind of work is less conspicuous to us than the mechanical work of contraction or the biosynthetic work involved in cell growth, but it is of equal importance in cell function. All cells are capable of accumulating certain essential substances from the environment, either minerals such as potassium, or nutrients such as glucose, so that their intracellular concentration may be much higher than in the medium outside the cell. Conversely, unwanted or deleterious substances may be actively pumped out of the cell, or *secreted,* even when the external concentration of the substance is much higher than the internal. Such movements of molecules against gradients of concentration cannot occur spontaneously, since solute molecules normally tend to distribute themselves in all the space available to them, so that they are completely randomized in it. The term *active transport* is applied to the energy-dependent movement of solute molecules against their tendency to randomize. Through the action of active transport "pumps" in their membranes, cells can maintain

TABLE 1–1. Chemical Work of Biosynthesis

Macromolecule	Building Block	Type of Bond	$+\Delta G^{0\prime}$ per Bond kcal/mole
Protein	Amino acid 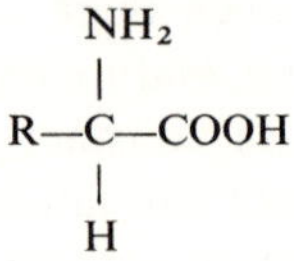 R—CH(NH_2)—COOH	Peptide —C(=O)—NH— 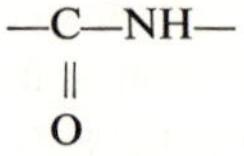	~5.0
Nucleic acid	Mononucleotide 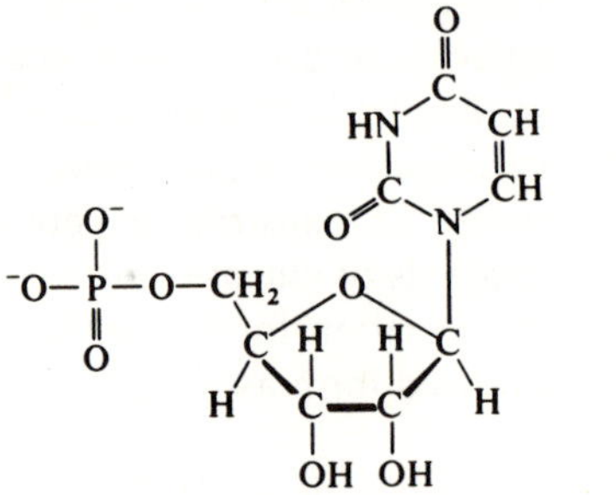	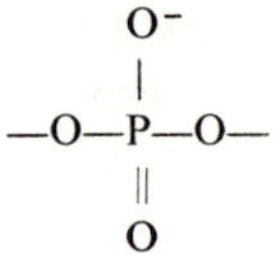 Phosphodiester —O—P(O^-)(=O)—O—	~5.0
Polysaccharide	Monosaccharide 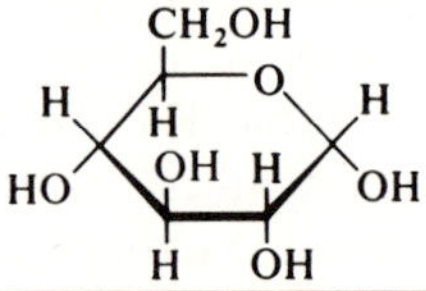	Glycosidic 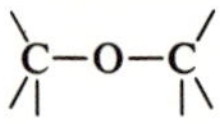 >C—O—C<	~3.0

their internal milieu constant and optimal for life even though the external environment may have a very different chemical composition. Furthermore, through active transport mechanisms, cells can also extract vital nutrient molecules from the environment, even when they are present in exceedingly low concentrations. The electrical activity of many cells is also the result of osmotic work, which is instrumental in the mechanism of excitation and the conduction of impulses in nerve and muscle cells.

Finally, it is a matter of common observation that most organisms can carry out mechanical work. Most conspicuous is the work done by the contraction of skeletal muscle in higher animals, which can be easily observed and measured. However, such conspicuous contractile processes are but refinements of a more generalized property of nearly all cells to exert intracellular pulling forces by means of contractile filaments. For example, during division of higher cells, contractile fibers in the cell are responsible for pulling apart the chromosomes in the nucleus and for dividing the cytoplasmic material. Motile structures such as *cilia* and *flagella* also perform mechanical work, that of propulsion. It is remarkable that the mechanical work done by living organisms is directly powered by chemical energy, whereas the man-made devices for doing mechanical work with which we are familiar run on heat or electrical energy.

All three types of work carried out by living organisms ultimately lead to dissipation of energy and its ultimate randomization in the environment. Because there is friction at each of the many sequential steps in biological energy conversion, a large fraction of the energy originally captured from sunlight by the green plant cell is lost to the environment as heat. Even the final performance of cell work ultimately results in dissipation of energy. For example, man does mechanical work on his environment when he lifts stones to build a wall; in time, however, the wall will crumble and its components will randomize in the environment. The work done in biosynthesis and in maintaining the intracellular electrolytes is also ultimately dissipated when cells die and their contents are dispersed in the environment. In the flow of energy in the biological world, then, we have an inevitable and irreversible degradation of energy. The high-grade useful energy of sunlight becomes the "medium-grade" chemical energy of organic molecules, and this in turn is ultimately dissipated and randomized in the environment. The flow of energy in the biological world is unidirectional and irreversible, because once energy is randomized, it can never again do biological work.

1–5 THE CYCLING OF MATTER IN THE BIOLOGICAL WORLD

Accompanying the flow of energy through the biological world there is a flow of matter (see Fig. 1–2). During respiration, animal cells take oxygen and organic nutrients from their surroundings and then discharge carbon dioxide and water. In contrast, green plants extract carbon dioxide and water from their surroundings, from which they make new cell material, and then return oxygen to the atmosphere. Moreover, there is also a constant cycling of available nitrogen. Plants obtain nitrogen from the soil as nitrates, which they convert into ammonia and then into amino acids. Amino acids generated by plants are consumed by animals and converted again into ammonia, which returns to the soil and is converted into nitrates by nitrifying bacteria. Carbon dioxide, oxygen, nitrogen, and water thus cycle continuously between the animal and plant worlds. The pool of available water on the earth is very large, as is the content of oxygen in the atmosphere. However, the pool of available carbon dioxide is rather small; the atmosphere contains only about 0.03 per cent carbon dioxide. In fact, it has been estimated that if all the combustion processes producing carbon dioxide were suddenly halted, the

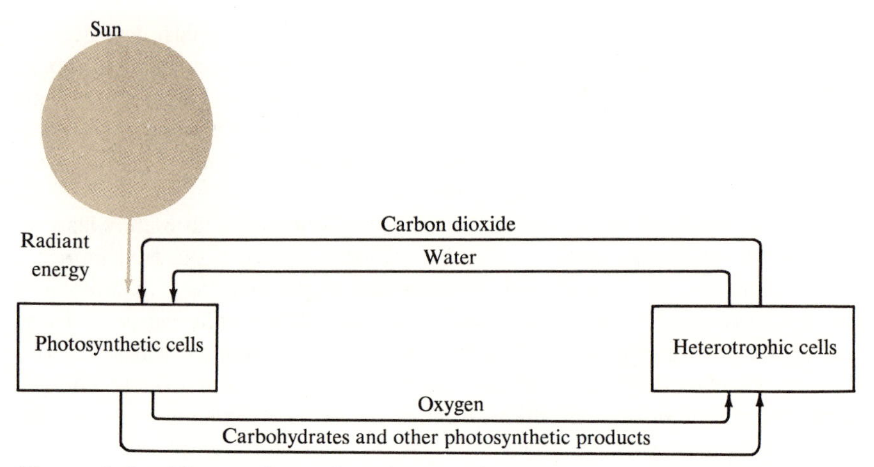

Figure 1–2. The cycling of carbon and oxygen through the plant and animal worlds.

plant life of the earth would consume all the available atmospheric carbon dioxide in a year or two. There is thus a fine balance between carbon dioxide production and utilization in the biosphere. The world of photosynthesizing cells and the world of heterotrophic cells live in symbiosis with each other, a symbiosis in which both matter and energy are component elements.

1–6 THE IMMENSITY OF THE BIOLOGICAL ENERGY CYCLE

From the amounts of carbon dioxide utilized by photosynthesis and from the amounts of energy required to convert carbon dioxide into glucose, it is possible to estimate the magnitude of the energy flow in the biological world. The total amount of carbon fixed by all the photosynthetic organisms on the face of the earth per year is thought to be between 130 and 160×10^9 tons, the great bulk of which is accounted for by marine plants (Table 1–2). Actually some authorities feel that this estimate should be much higher because photosynthesis in marine plants may be very grossly underestimated. However, let us take these figures and assume that all the fixed carbon is in the form of glucose and that the formation of each mole of glucose (180 g) requires input of 686 kcal of solar energy. We will see that this amounts to a biological energy flux of about 1×10^{18} cal/yr. However, on the average only about 2 per cent of the light impinging on a plant is actually converted into the chemical energy of glucose. Therefore, the total amount of solar energy absorbed by the plant world is about 5×10^{19} cal/year. This in turn is but a very small fraction (1/2,000) of the total solar energy falling on the profile of the earth each year, which is estimated to be about 1×10^{24} cal/yr.

TABLE 1–2. Annual Fixation of Carbon by Photosynthesis

Region	Area km²	Carbon Fixed tons/km²/yr	Total Carbon Fixed tons/yr
1. Forest	44×10^6	250	11×10^9
2. Cultivated land	27×10^6	160	4.3×10^9
3. Grassland	31×10^6	36	1.0×10^9
4. Desert	47×10^6	7	0.3×10^9
Total land	149×10^6		16.6×10^9
Total ocean	361×10^6	340	122.6×10^9

Now let us compare the figure for the magnitude of the biological energy flow with the amount of energy estimated to be expended each year by all the man-made machines on the face of the earth. This figure is difficult to fix exactly because it is evidently increasing with time, but at present it is believed to represent no more than about 3 per cent of the annual flux of biological energy. Moreover, most man-made machines burn biologically formed fuels, such as coal, oil, and natural gas, derived from fossil plant material laid down in prehistoric times. Nonbiological sources of energy, such as water power and nuclear power, still furnish only a small fraction of the energy requirements in all man-made machines.

1–7 THE PATTERN OF ENERGY FLOW IN CELLS

Now let us turn from the biological macrocosm to the biological microcosm and sketch the pattern of energy flow within single cells, the smallest units of living matter. All cells, whether animal, plant, or microbial, utilize the same fundamental molecular principles and mechanisms in their energy-transforming activities. We can therefore use the more familiar heterotrophic or "animal" cell as the basis for our discussion.

The energy-transforming activities in the cell may best be visualized in terms of a flow of chemical energy from energy-rich foodstuff molecules to those energy-requiring or endergonic processes which are necessary for the function and survival of living cells.

The primary process in all heterotrophic cells for delivering useful chemical energy is the oxidative degradation of organic nutrient molecules such as glucose. Now we must define some terms. *Oxidation* is defined chemically as the loss of electrons from an atom or molecule; *reduction* is defined as a gain of electrons. Whenever we have an oxidation some other compound must undergo reduction. Electron donors are known as reducing agents and electron acceptors as oxidizing agents. The major electron donors in heterotrophs are their organic fuels, such as glucose. Some heterotrophs use molecular oxygen as their electron acceptor or oxidizing agent; they are called *aerobic* cells or *aerobes*. However, there are other heterotrophic cells, among them many bacteria, which do not use oxygen at all, or are even poisoned by it. Such cells are called *anaerobic* organisms or, more simply, *anaerobes*. What, then, do anaerobes use as oxidation agents for glucose? Instead of using oxygen, anaerobes have developed a remarkable trick. Most of them are able to break down the 6-carbon glucose molecule into two or more fragments. One of these fragments, which we may regard as the electron donor or reducing agent, now becomes oxidized by another fragment, which is the electron acceptor or oxidizing agent. This process by which foodstuff molecules undergo oxidoreduction in the absence of oxygen is called *fermentation*. In all cells, then, it is the process of oxidation or electron transfer which is the main source of useful chemical energy. In aerobic cells the oxidant is oxygen and in anaerobic cells it may be an organic substance derived from part of the nutrient molecule itself.

In both aerobic and anaerobic cells, the energy of the foodstuff molecule is conserved during its oxidation, not as heat, but as chemical energy. We can now be quite specific and say that the energy of cellular oxidation reactions is conserved in the compound *adenosine triphosphate*, which has been known universally to all biologists for over a generation by its initials: ATP. ATP is the carrier of chemical energy from the energy-yielding oxidation of foodstuff molecules, whether aerobic or anaerobic, to those processes or reactions of the cell that do not occur spontaneously and can proceed only if chemical energy is supplied (see Fig. 1–3). During the energy-yielding oxidation of foodstuffs in the cell, ATP is formed from adenosine diphosphate (ADP) in coupled reactions; some of the energy of the foodstuff molecule is thus saved or conserved as the energy of the newly formed ATP. The chemical energy of ATP is then used to perform the chemical, mechanical, and osmotic work of the cell, during which the terminal phosphate group of ATP is lost and ADP is formed. In brief, ATP is the "charged" form of the energy transporting system, and ADP the "discharged" form. The many chemical steps in the charging and discharging of the ATP system in the cell are catalyzed by enzyme systems. This is the basic principle of the cellular energy cycle, and central to it is the substance ATP.

1–8 SPECIALIZED ENERGY CONVERSIONS

In addition to the mainstreams of biological energy flow that we have described, living organisms have perfected other interesting means of transforming energy. For example, the electric eel can deliver a shock of several

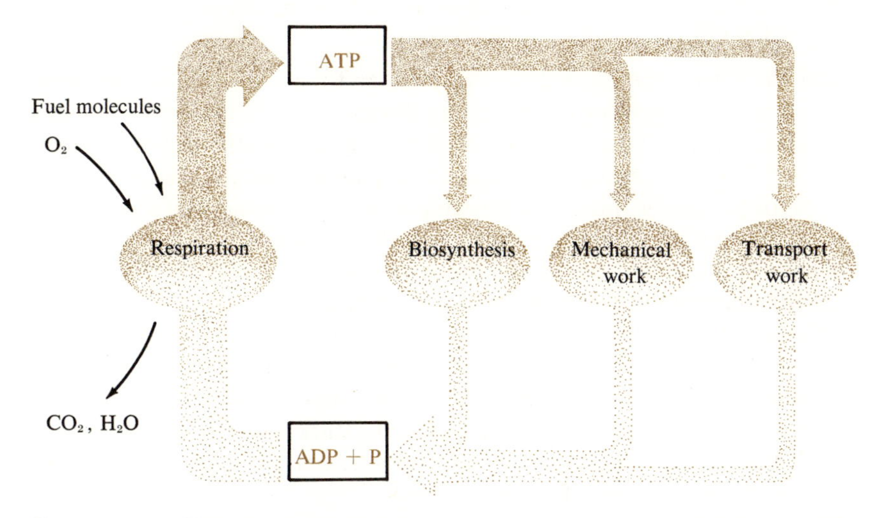

Figure 1–3. The transfer of energy in heterotrophic cells by the ATP-ADP system.

hundred volts of electrical potential, which is derived from chemical energy. The flashes of the firefly are brought about by the conversion of chemical energy into light. The bombardier beetle can fire its startling cannon by converting the chemical energy of hydrogen peroxide into the pressure-volume energy of expanding oxygen gas.

Living organisms also possess energy-sensing devices of incredible sensitivity. The human ear can sense the most delicate differences in the timbre and overtones of musical notes, although only minute amounts of sonic energy perturb the air. The human eye is responsive to extremely small gradations of light energy. Biological energy-sensing devices often are far superior in sensitivity and efficiency to man-made instruments, even in this electronic age.

There are still other ways, less obvious and more subtle, in which energy is utilized and transformed by living organisms. For example, energy is required to create the very complexity of form of a living organism and the great morphological diversity of different species of life. Living organisms are rich in information, which can be regarded as a form of energy. As we shall see, it is a fundamental law of thermodynamics that all atoms and molecules in the universe inexorably tend to seek the most random or disordered state, with the least energy content, that is, the state of maximum entropy. Thus the maintenance of biological complexity and its ongoing evolution toward further complexity are matters of the most fundamental and far-reaching theoretical concern to biologists.

1–9 CELL STRUCTURE AND THE DIVISION OF LABOR IN CELLS

A cell has often been likened to a factory, and the similarity is even more compelling in the light of the preceding discussion. A large factory consists of a series of divisions, and each of these contains a number of separate machines linked according to some master plan so that the whole complex can produce specific products through a series of sequential operations. Similarly, living cells contain specific intracellular structures, each of which has a definite function and role in the organization of cellular activity.

Considered from the point of view of their internal structure, all cells may be placed into one of two great classes: *prokaryotic* and *eukaryotic cells*, which are illustrated in Figs. 1–4 and 1–5, respectively. These terms are derived from *karyon*, which means kernel, or nucleus. Prokaryotic cells (*pro* means before or preceding) possess no nucleus, but rather a structure that may be regarded as a precursor of a nucleus. Eukaryotic cells (*eu* means well), on the other hand, possess a well-formed nucleus.

Prokaryotic cells include the bacteria and the blue-green algae. They are much simpler in structure than eukaryotic cells. Prokaryotes lack not only a nucleus but also many other internal structures found in eukaryotes, such as mitochondria and endoplasmic reticulum. Moreover they are relatively very

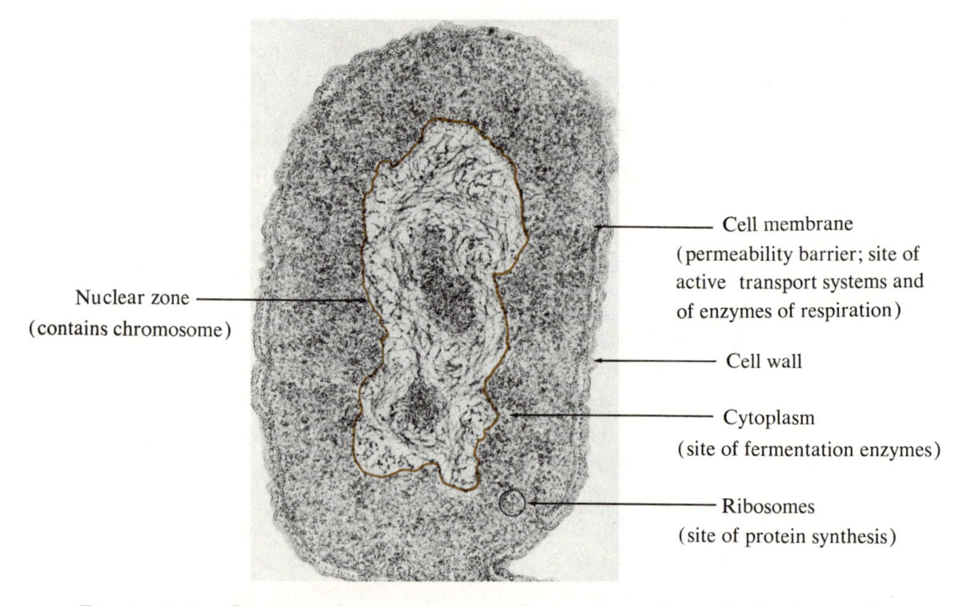

Figure 1–4. Structural organization of a prokaryotic cell, the bacterium *Escherichia coli*. Electron micrograph by courtesy of Mr. Glenn Decker, Johns Hopkins University.

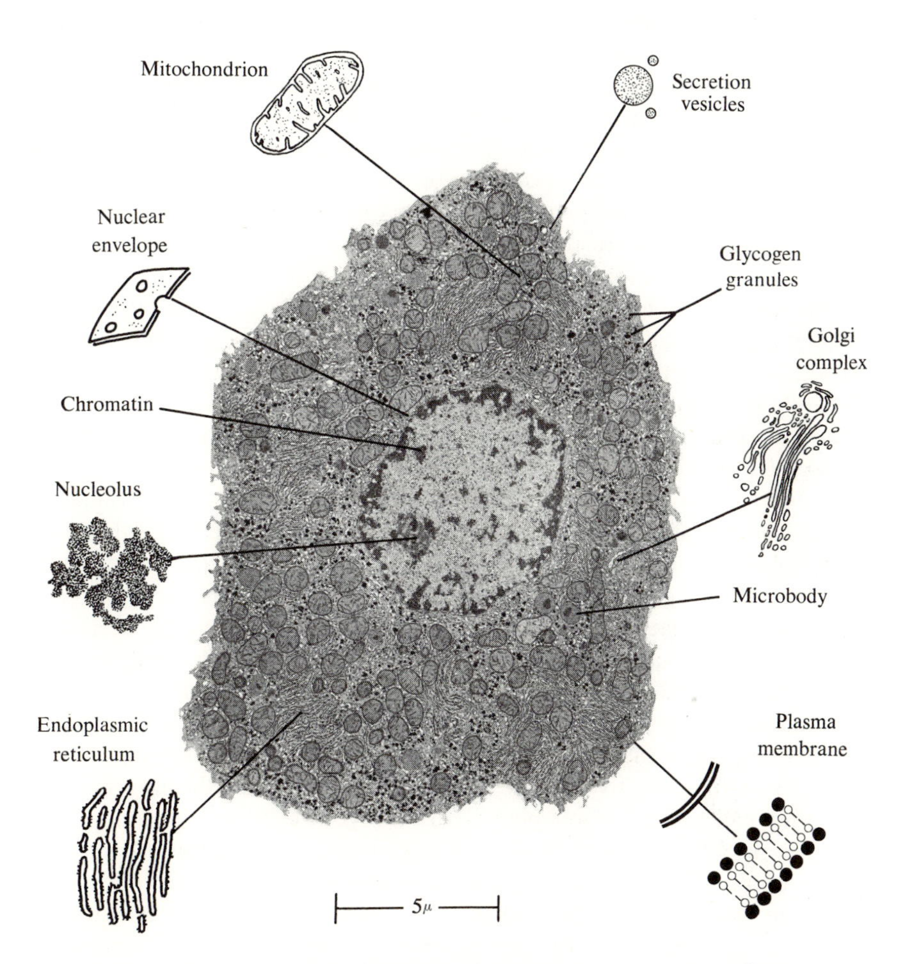

Figure 1–5. Structural organization of a typical eukaryotic cell, a parenchymal cell of rat liver. Electron micrograph by courtesy of Mr. Glenn Decker. Drawings adapted from W. Bloom and D. W. Fawcett, *A Textbook of Histology*. Philadelphia: W. B. Saunders Co., 1962, p. 21.

small; a typical bacterial cell is only about 1–2 μ in length and has only about 1/5000 of the volume of a typical eukaryotic cell, such as a liver cell of a vertebrate. Despite their conspicuous structural differences prokaryotic and eukaryotic cells are constructed of the same kinds of building-block molecules —sugars, fatty acids, and amino acids, for example—and they carry out similar kinds of enzymatic and metabolic reactions. Prokaryotes contain only one chromosome, a single molecule of deoxyribonucleic acid or DNA.

Because prokaryotic cells are very simple, primitive cells we might think they are evolutionary curiosities unworthy of our attention. But we must recall that because of their sheer abundance on the face of the earth, prokaryotic microorganisms are responsible for a very large fraction of the total biological energy flux in the biosphere.

The great class of eukaryotic cells includes a number of unicellular organisms other than bacteria, such as yeasts, molds, protozoa, and higher algae, as well as the cells of nearly all multicellular organisms or *metazoa.* All eukaryotic cells contain the same basic intracellular structures—nucleus, mitochondria, and endoplasmic reticulum. However, in contrast to prokaryotes, eukaryotic cells are far more differentiated and specialized.

Figure 1–5 shows an electron micrograph of a eukaryotic cell. Surrounding it are drawings showing the structure of the major organelles or subcellular morphological units as they are magnified by the electron microscope. Also indicated are the probable sites of the different types of energy conversion. It is noteworthy that each of the important transformations of energy in eukaryotic cells takes place within a specific, well-defined intracellular organelle. For example, we see that the mitochondria, small membrane-surrounded structures in the cytoplasm, are the sites of the enzyme systems that participate in the oxidation of foodstuffs by molecular oxygen and the recovery of the oxidative energy as ATP. For this reason the mitochondria are often called the "power plants" of eukaryotic cells. In prokaryotic cells the enzymes for oxidation of nutrients are located in the cell membrane.

The energy of ATP is then utilized in other parts of the eukaryotic cell to carry out work. For example, much of the osmotic work of the cell actually takes place in the cell membrane, in which are located enzymes capable of carrying out the directional work of active transport. The ribosomes, very small granular bodies found in both prokaryotic and eukaryotic cells, are the site of protein synthesis; they are literally protein factories. The endoplasmic reticulum of eukaryotic cells contains important enzymes required in biosynthesis of lipids. The cell nucleus contains the genetic material in the form of DNA, which is divided into several or many chromosomes. The nucleus is the site of the replication of DNA and its transcription to form ribonucleic acid (RNA), which are processes that also require the energy of ATP. Finally, eukaryotic cells contain contractile filaments that are sometimes highly specialized and differentiated, as in the myofibrils of skeletal muscle.

Each of the intracellular structures or organelles has a very complex ultrastructure of its own. We shall see later that the groups of enzyme molecules that cooperate to bring about energy transformations in these organelles are arranged in characteristic supramolecular assemblies. In fact, study of the molecular organization of cell organelles such as mitochondria, chloroplasts, and ribosomes is now one of the most challenging fields of biochemistry and molecular biology.

We have now seen that all living cells are endowed with extremely effective devices for transforming energy. A single cell of microscopic dimensions may contain many different kinds of energy-transforming systems and many sets of each type. The ultimate units in the organization and function of cellular energy-transforming systems are single molecules of specialized proteins. Actually, many of these molecular devices are the envy of the industrial engineer. We are all aware that chemical energy can be converted into heat energy by combustion and that heat can be converted into mechanical energy by the steam engine. We also know as a matter of common experience that mechanical energy can be converted into electrical energy by a generator and that electrical energy can be converted into heat, as in an electric toaster, or into chemical energy, as in a battery charger. However, these examples by no means cover all of the energy conversions that are theoretically possible; they simply happen to be the easiest ones for the engineer to employ. Today, we are confronted in our modern technology with a requirement for new modes of energy conversion for special purposes, such as solar batteries, thermoelectric converters, and fuel cells, in which chemical energy may be directly converted into electrical energy.

Living organisms are today providing the engineer with models and ideas for new methods of converting energy, with the hope that efficient man-made counterparts for some of the known biological "machines" can be developed. We have already mentioned the conversion of solar energy into chemical energy by the green plant, for which no man-made counterpart exists. To date no cheap, efficient process for desalination of sea water has been perfected, yet every living cell is equipped with molecular devices in its membranes that can pump salts and water with extraordinary efficiency. It is also extraordinary that one of the most conspicuous energy conversions in the living organism, namely the direct conversion of chemical energy into mechanical energy, such as occurs in the contraction of muscle, is scarcely applied at all in our industrial technology. Mechanochemical engines might one day play a very useful role in technology if they could be developed to be as efficient as skeletal muscle.

Finally it may be pointed out that biological evolution has not only selected remarkably efficient devices for absorbing, converting, and sensing energy, but it has selected them at the molecular level. In this day when miniaturization of electronic and computing equipment is a major goal in our technology, the ultimate in miniaturization has already been achieved by the molecular machines in living cells.

2

PRINCIPLES OF THERMODYNAMICS

To beginning students, thermodynamics often appears to be rather abstract and formidable. Its principles are developed with the use of idealized model systems, which are subjected to imaginary manipulations of variables. Moreover, its derivations appear to be arid exercises in calculus. However, the working philosophy and approaches of thermodynamics are simple and logical and require no special knowledge of chemistry or molecular theory. Furthermore, we need to become familiar with only a small part of thermodynamic formalism in order to examine in a broad way the nature of biological energy transformations.

2–1 THE SCOPE OF THERMODYNAMICS AND ITS APPROACHES

Thermodynamics is perhaps the most fundamental and exact field of physics and it provides one of the most general methods for analyzing physical and chemical phenomena. It is a statistical science and the accuracy of its predictions increases with the size of the body of matter under study. Furthermore, it deals with relatively gross or macroscopic properties of matter in the bulk, such as pressure, temperature, volume, and chemical composition, properties that are easily measured with simple methods. The analysis of chemical processes by thermodynamics requires no detailed knowledge of atomic theory or the mechanism of reactions; in fact, it is independent of any conception we may have of the structure of matter.

Exactly how does the science of thermodynamics proceed to analyze energy exchanges? First, we must specify the collection of matter in which we wish to study the energy changes during some chemical or physical process. This collection of matter is called the *system*; all the other matter in the universe apart from this system is called the *surroundings*. Next, we must specify the total energy content of the system before the process has taken place, called the initial state, and then again after the given process has taken place, that is, the final state. As the system proceeds from its initial to its equilibrium state, it may either absorb energy from or deliver energy to the surroundings. The difference in the energy content of the system as it passes from its initial to its final equilibrium state is counterbalanced by a corresponding and inverse change in the energy content of the surroundings. The equilibrium state is that condition in which no further change is occurring within the system, or between the system and its surroundings. At equilibrium the temperature and pressure are uniform throughout the system and no unbalanced forces are still operating.

We have just said that we must specify the total energy content of the initial and final states of the system under consideration. However in practice it is usually a formidable and often an impossible task to measure accurately the total energy content of any given system. But often we can easily measure the *difference* in energy content between the initial and equilibrium state of the system, that is, the amount of energy exchanged between the system and its surroundings as the process occurs. If we know both the amount and the kind of energy a given system exchanges with its surroundings as it passes from the initial state to the equilibrium state, we will have enough information to carry out thermodynamic analysis of a process.

Thermodynamic laws are independent of the time that is required for a system to pass from the initial to the final state. Thus, it does not matter whether a given chemical or physical change requires seconds or centuries to come to equilibrium. Moreover, thermodynamics does not concern itself with the pathway or mechanism of physical or chemical changes. It is concerned only with the energy difference between the initial state and the final state of the system, regardless of how this transition occurred. An analogy will clarify this point. If a person travels from New York to San Francisco, his change of location depends only on the difference between the latitude and longitude of his initial position (New York) and the latitude and longitude of his final position (San Francisco), no matter which of a thousand possible routes he may take or whether he travels three thousand or ten thousand or one million miles to get there. So it is in thermodynamics. Only the initial and final states matter; the pathway taken by the process is irrelevant. By direct corollary, the energy changes occurring in each of the intermediate steps in any chemical or physical process are completely additive. Their algebraic sum is exactly equal to the energy change of the overall process, regardless of the nature or number of steps in the pathway.

When the thermodynamicist studies energy changes during a process, he tries to reduce to a minimum the number of variables affecting the system, such as pressure, volume, or temperature. For example, if the temperature of a system is held constant, it becomes easier to specify exactly the changes in energy which occur in going from the initial to the final state. In the simple biochemical reactions we wish to scrutinize, we can achieve a very great simplification because biological reactions usually take place in dilute aqueous solutions that are in equilibrium with the atmosphere, conditions in which not just one but three of the most important variables, namely, temperature, pressure, and volume, all remain constant. For this reason we need not concern ourselves with the large part of thermodynamic formalism that deals with gases, pressure-volume changes, and transitions between different states of matter.

2–2 THE FIRST LAW OF THERMODYNAMICS

All events in the physical world conform to and are determined by the two fundamental principles of thermodynamics, known as the *First Law* and the *Second Law*.

The First Law, enunciated by Robert Mayer in 1841, is the principle of the conservation of energy: *energy can be neither created nor destroyed.* Thus in any chemical or physical process the total energy of the system plus surroundings, that is, the total energy of the universe, remains constant.

We already know that different forms of energy may be interconverted. Thus, the thermal energy of steam can be transformed into mechanical energy by a steam engine. Mechanical energy can be converted into electrical energy by a generator and electrical energy can in turn be converted into chemical energy, as in the charging of a storage battery. The First Law of thermodynamics imposes only one limitation on such energy exchanges, namely, that the total energy of the system plus surroundings must remain constant.

The First Law also implies that there is a quantitative correspondence between different kinds of energy. From many physical measurements such quantitative energy equivalences have been established. For example, we may recall that there is a mechanical equivalent of heat. Mechanical energy is usually measured in terms of the erg, the energy delivered by a force of 1.0 dyne acting through a distance of 1.0 cm. Heat energy, on the other hand, is expressed in terms of the *gram calorie*, which we have already defined (Section 1–2), or the joule (1 cal = 4.1855 joules). From many exact physical measurements it has been deduced that 1.0 gram calorie of heat energy is equivalent to 4.185×10^7 ergs of mechanical energy, and vice versa, 1.0 erg of mechanical energy is equivalent to 2.389×10^{-8} gram calories or 1×10^{-7} joules. Similarly it has been found that 1.0 gram calorie of heat energy is theoretically equivalent to 4.185 watts of electrical energy. The existence of such equivalences between different forms of energy implies that they are easily and completely interconvertible; however, this is not always the case. For example, mechanical energy can be quantitatively converted into heat energy in an appropriate apparatus, but the reverse process, the conversion of heat energy into mechanical energy, is never complete under normal circumstances.

Because heat is the most familiar form of energy and is easily measured, most early investigations of the equivalence of different forms of energy and of the energy exchanges occurring in physical and chemical processes were evaluated on the basis of heat changes. For this historical reason the science dealing with energy exchanges was called *thermo*dynamics. But this science actually deals with and is relevant to exchanges of *all* types of energy; a more accurate name for it might be *energetics*.

2–3 HEAT CHANGES IN CHEMICAL REACTIONS

Virtually every physical or chemical event is accompanied by absorption of heat from or release of heat to the surroundings. When a process occurs with loss of heat to the surroundings, it is called *exothermic*; when it occurs with absorption of heat from the surroundings, it is called *endothermic*. Heat is therefore a simple and universal medium or form in which energy can be transferred.

The heat changes occurring during chemical reactions are easily measured in a calorimeter and can give us valuable information about the amount of work a chemical reaction can perform under certain conditions. In order to simplify and systematize calculations of heat and energy exchanges of chemical reactions, they are always given in terms of kilocalories of energy exchanged per gram molecular weight of compound transformed. Moreover, the energy exchanges occurring in chemical reactions always assume that all the reactants and the products are present in their *standard states*. The standard state of a substance is defined as its most stable form at 1.0 atmosphere pressure at 25°C (298°K). For biological reactions occurring in aqueous solution, the standard state of a reactant or product is defined as a concentration of 1 gram molecular weight per liter of water; the hydrogen ion concentration in the standard state is taken as 1.0×10^{-7} M or pH 7.0.

In Table 2–1 we see that organic compounds have a characteristic *heat of combustion*, which may be defined as the number of kilocalories of heat given up to the surroundings as 1 mole or gram molecular weight of a substance is

TABLE 2–1. Heats of Combustion of the Major Fuel Molecules

Fuel	Molecular Weight	Heat of Combustion kcal/mole	kcal/g
D-glucose, a carbohydrate	180	673	3.74
Tripalmitin, a fat	809	7510	9.3
Glycine, an amino acid	75	234	3.12

burned completely at the expense of molecular oxygen. For example, the equation for combustion of glucose is

$$C_6H_{12}O_6 + 6O_2 \longrightarrow 6CO_2 + 6H_2O$$

The heat given off by the oxidation of 1 mole of glucose (180 g) under the standard conditions defined above is 673 kcal. Glucose produces heat during combustion because its complex structure possesses much potential energy, which is released when glucose is degraded to the simpler products CO_2 and H_2O. Table 2–1 also shows the heat of combustion of other important organic fuels of cells such as fats and amino acids. In general, combustion of fat yields much more energy per gram than combustion of either carbohydrates or

TABLE 2–2. Heat Changes of Some Representative Chemical Reactions, Under Conditions of Constant Temperature and Pressure

Type	Heat Change kcal/mole
Oxidation	
glucose + $6O_2 \rightarrow 6CO_2 + 6H_2O$	—673
Hydrolysis	
sucrose + $H_2O \rightarrow$ glucose + fructose	−4.8
glucose 6-phosphate + $H_2O \rightarrow$ glucose + H_3PO_4	−3.0
Neutralization	
$NaOH + HCl \rightarrow NaCl + H_2O$	−13.8

amino acids. Fat is a more concentrated fuel than glucose because it is more fully reduced or hydrogenated, and thus requires more oxygen to burn it to CO_2.

Table 2–2 shows the heat changes of other types of chemical reactions. We see that hydrolytic reactions are also exothermic, but they proceed with much less evolution of heat than do combustions. The neutralization of H^+ and OH^- ions is also an exothermic process. However, in other types of chemical reactions, such as certain ionizations, heat may actually be absorbed from the surroundings; such reactions are thus endothermic. Heats of reactions can be determined not only calorimetrically but can also be predicted from the known heats of similar or related reactions.

2–4 THE ISOTHERMAL NATURE OF CELLULAR PROCESSES

We now come to a striking difference between living cells and man-made energy-converting devices. Although heat is the simplest and most familiar medium by which energy may be transferred or used in man-made machines, it is not a useful form of energy for performing biological work. Why is this so?

Heat can do work only if there is a temperature differential through which it can act. Heat must either pass from one body to a second one having a lower temperature, or the temperature of a given body containing the heat energy must be lowered. However, it is a matter of common observation that there are no great temperature differences between one part of a living organism and another.

Yet is it not possible that there are very small temperature differentials in cells that might serve as the basis for doing cellular work? We can provide a quantitative answer to this question. The maximum work w that may be derived from a heat engine is given by the equation

$$w = q\left(\frac{T_2 - T_1}{T_2}\right)$$

where q is the heat absorbed and T_2 and T_1 are the absolute temperatures (°K) of the bodies of matter between which the heat passes. For example, T_2 could be the temperature of steam entering the piston of a steam engine, and T_1 the temperature of the exhaust steam after the expansion stroke. It is clear from this equation that the maximum work can be derived from heat only if the temperature of the exhaust steam is absolute zero. We can now see the basic restriction that is placed on the conversion of heat into work. If in a system at ordinary temperature of, let us say 25°C (298°K), there is only a very small temperature differential (0.1°C), the fraction of a given amount of heat energy that can be converted into work in this system is trivially small. For all practical purposes, then, living organisms are isothermal and thus cannot function as heat engines.

In order to understand how cells perform work under isothermal conditions, we must define another form of energy, called *free energy*. But first we must examine the Second Law of thermodynamics, from which the concept of free energy emerges.

2–5 ENTROPY AND THE SECOND LAW

The First Law tells us that energy is conserved; every physical or chemical change must satisfy this principle. However, there is another fundamental aspect of energy exchanges not explained by the First Law. A simple example will serve to illustrate the problem (Fig. 2–1).

Suppose we place two blocks of copper together, one hot and one cold, seal them in an insulated container, and allow them to come into equilibrium with each other. The temperature of the hot block will decrease and that of the cold block will increase until they both reach some intermediate equilibrium temperature, which will be uniform throughout both blocks. The flow of heat and thus of energy from the hot block to the cold one is spontaneous. However, if we put two identical blocks of copper, both at the same temperature, into such an insulated container, we know that they will remain at the same temperature. We would never expect the temperature of one block to rise spontaneously and that of the other to fall. However, even if this should happen, it would not violate the First Law because the energy lost by one block might be gained by the other; the total energy of the two blocks would still remain the same.

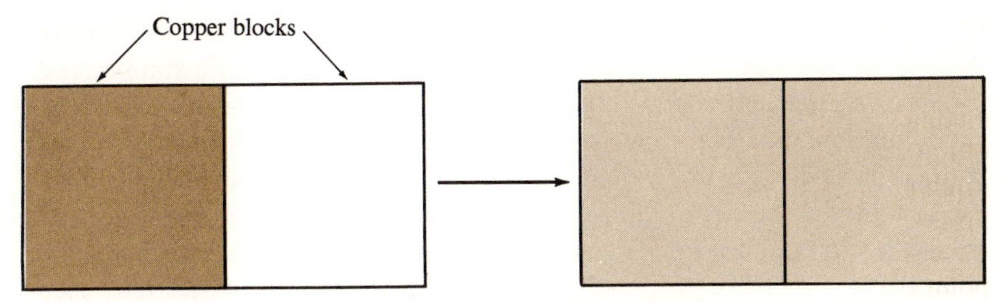

Heat flows from warm body to cool body

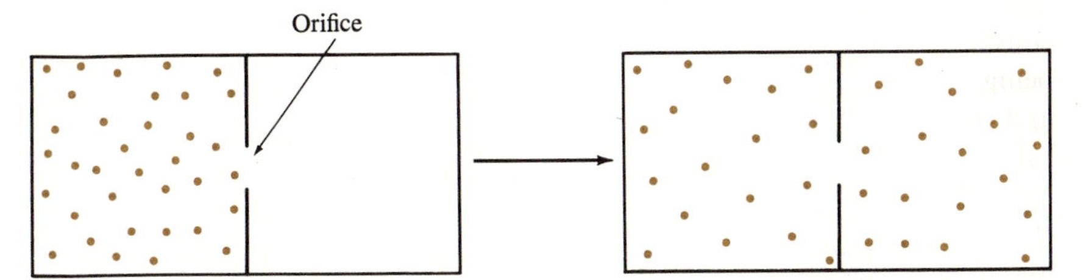

Gas molecules flow from a zone of high pressure to one of low pressure

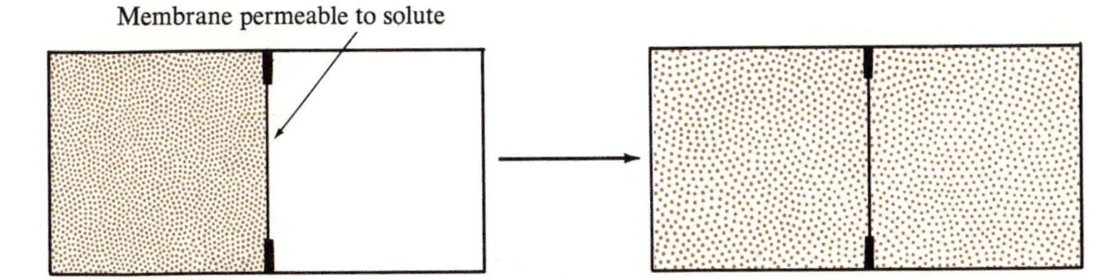

Solute molecules flow from a zone of high concentration to one of low concentration

Figure 2–1. The increase of entropy or randomness in some physical systems. Such flows *never* reverse spontaneously.

It is quite clear from considering this example (and others shown in Fig. 2–1) that spontaneous physical or chemical changes have a *direction* that cannot be explained by the First Law. We have seen that all processes proceed toward an equilibrium state, in which the temperature, pressure, and all other measurable attributes have become uniform throughout the system and surroundings. However, once a process has reached such an equilibrium it will not reverse spontaneously to the initial state. When the hot and cold blocks of copper in our example reach exactly the same temperature, all the heat energy originally contained in the two blocks is uniformly distributed throughout the blocks. The energy has become maximally randomized and will never by itself "unrandomize."

The Second Law of thermodynamics provides us with a criterion for predicting the direction of any given process. First, it recognizes a state or condition of matter and energy called *entropy*, which may be simply defined as randomness or disorder. Then it states that all physical and chemical processes proceed in such a direction that the randomness or entropy of the universe—the system plus its surroundings—increases to the maximum possible; at this point there is equilibrium. The Second Law says that no process can occur in such a way that the entropy of the universe decreases. Processes that occur with an increase in entropy are thus irreversible; they will never return to their initial state. Theoretically, the entropy of the universe may remain constant during a process, and when it does such a process is spoken of as being *reversible*. Although we can therefore distinguish between reversible and irreversible processes, we must add at once that completely reversible processes in which entropy remains exactly constant are hypothetical; they are never seen in our real physical world. Why is this so? Real processes are irreversible because one form of energy cannot be quantitatively converted into another without some loss of energy in a randomized or dissipated form. For example, the mechanical energy applied to an electrical generator is not completely converted into electrical energy because of energy lost in friction, which is dissipated as heat into the environment, in which it randomizes and thus becomes unavailable to do work. All the matter and energy in the universe are thus undergoing constant randomization. It is the ultimate destiny of the universe, then, to attain a state of complete randomness and disorder, which has been called *entropic doom*.

If entropy is randomness or disorder, its opposite is order. We may go ahead now and state a relationship which is sometimes called the Third Law of thermodynamics: the entropy of a perfect crystal of any element or compound at absolute zero temperature is zero. Because there is no thermal motion at absolute zero, the atoms of a perfect crystal would then be in absolutely perfect order. Any state less orderly than such a crystal has a finite amount of entropy or randomness.

Entropy is expressed in terms of *entropy units*, which have the dimensions of calories per mole per degree. At any given temperature, solids have relatively low entropy, liquids an intermediate amount, and gases the highest entropy; the gaseous state is the most disordered and chaotic state of matter. Moreover, gas at a high temperature has more entropy than at a low temperature because of the greater thermal motion of its molecules. For this reason, the temperature must be given when we wish to specify entropy changes.

2–6 FREE ENERGY

We have just seen that the driving force in all physical and chemical processes is the tendency for a system to seek that state in which the entropy of the universe is maximized. But entropy or entropy changes are not always easily measured or calculated. However, under special conditions, the change in entropy during a process is quantitatively related to changes in the total energy of a system by a third function called the *free energy*. This relationship is important in biochemical energetics because changes in free energy are easily measured and can be used to predict the direction and equilibrium state of chemical reactions.

Entropy and free energy are related by an equation that combines the First and Second Laws. This equation assumes that temperature, pressure, and volume are constant, and it is applicable only under these conditions. It is

$$\Delta G = \Delta E - T\Delta S$$

in which ΔG is the change in free energy of the system alone, ΔE is the change in total energy of the system, T is the absolute temperature, and ΔS is the change in entropy of the system plus surroundings (the universe). We use the symbol G for free energy in memory of Willard Gibbs, the American theorist who first developed this relationship. Because this equation holds for those conditions under which cellular reactions occur, that is, constant temperature, pressure, and volume, it is an important and fundamental relationship in biochemical energetics. If we now rearrange this equation to give

$$\Delta E = \Delta G + T\Delta S$$

we note that the change in total energy of the system (ΔE) is the algebraic sum of the term $T\Delta S$, which is always positive in any real process, and the term ΔG, which is always negative in any real process. The free energy change ΔG may now be defined as that fraction of the total energy change of a system that is available to do work as a system at constant temperature and pressure proceeds to the condition of equilibrium. Moreover, just as the entropy of the system plus surroundings proceeds to a maximum as a process goes to equilibrium, the free energy of the system alone proceeds to a minimum.

With the aid of Fig. 2–2 we may now summarize the relationships of the First and Second Laws when applied to systems at constant temperature and pressure. First we see that at all times the total energy of the system plus surroundings remains constant; however, we may note that the total energy of the system alone may increase, decrease, or stay constant. Secondly, the entropy of the system plus surroundings always increases to a maximum during a process. Thirdly, the free energy of the system alone decreases to a minimum. Because most reactions of biological interest are carried out at constant temperature and pressure, and because changes in free energy are easily measured, the direction and equilibria of biochemical reactions are

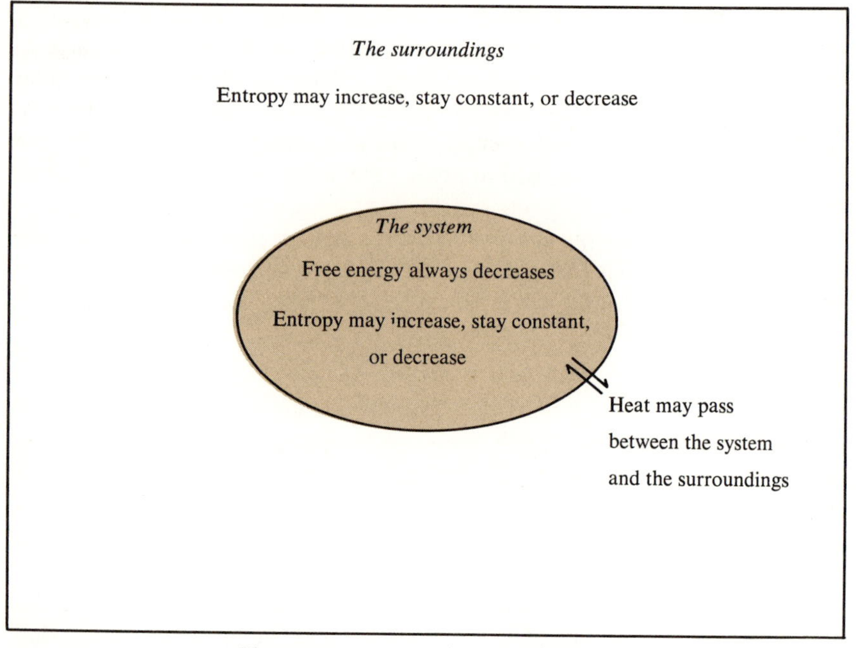

The First Law: The total energy of the universe is constant

The Second Law: The entropy of the universe always increases

Figure 2–2. The relationships between entropy and free energy changes in the system and surroundings in processes occurring at constant temperature and pressure.

normally predicted from their free energy changes rather than from their entropy changes.

We may note, as an important digression, that the entropy of the system alone does not necessarily increase during a process; it may increase, decrease, or stay constant. However, the Second Law says that the entropy of the *universe* must increase. Therefore, if the entropy of the system alone should decrease in a given process, the entropy of the surroundings must increase by such an amount that the *sum* of the entropy changes in the system *plus* surroundings increases. This is an extremely important point with regard to biological systems. When living organisms grow they obviously decrease in entropy, because of the highly structured nature of living matter. But this decrease in entropy can occur only if the surroundings increase in entropy. Put in another way, living organisms create their own internal order at the expense of the order of their environment, which they cause to become more random.

Now let us see how free energy changes can be measured.

2–7 FREE ENERGY AND THE EQUILIBRIUM CONSTANT

In a chemical process such as the reaction

$$A \longrightarrow B$$

a point of equilibrium will be reached at which no further net chemical change takes place; at this point the rate of conversion of A into B will then be exactly counterbalanced by the rate of conversion of B into A

$$A \rightleftharpoons B$$

There is a constant that expresses the chemical equilibrium reached by this system, the *equilibrium constant*, which has the form

$$K_{eq} = \frac{[B]}{[A]}$$

where the brackets represent the concentrations of the reactant and product that exist at the point of equilibrium. K_{eq} is an immutable constant at specified conditions of temperature and pressure, no matter what the starting concentrations of each component may have been at the beginning of the process. The equilibrium reached is the point at which the free energy of the system is minimum.

When the reaction has more components, the equilibrium constant is the product of the concentrations of all the reaction products divided by the product of the concentrations of all the reactants at equilibrium. For example, in the reaction

$$aA + bB \rightleftharpoons cC + dD$$

where a, b, c, and d are the number of molecules of A, B, C, and D respectively, the equilibrium constant is

$$K_{eq} = \frac{[C]^c\,[D]^d}{[A]^a\,[B]^b}$$

From everything we have said, it is obvious that the free energy change of a chemical reaction must be some mathematical function of its equilibrium constant. Actually, it is given by a very simple equation

$$\Delta G^{0\prime} = -RT \ln K'_{eq}$$

where R is the gas constant (1.987 cal/mole deg), T the absolute temperature, and ln K'_{eq} the natural logarithm of the equilibrium constant, at 25°C and pH 7.0. The symbol $\Delta G^{0\prime}$ designates the *standard free energy change*, the gain or loss of free energy in calories as 1 mole of reactant is converted to 1 mole of product at 25°C and pH = 7.0 under conditions where each is maintained in its standard state, that is, 1.0 molar for reactions in solution. The standard free energy change of a chemical reaction gives the maximum amount of work that the reaction can do at constant temperature and pressure. This amount of work can be realized only if some frictionless device exists by which the energy can be harnessed.

Table 2–3 shows the relationship between the equilibrium constant of chemical reactions and the calculated standard free energy change. When the equilibrium constant is greater than 1.0 (meaning that the reaction tends to go toward completion as written), then the standard free energy change is negative; such a reaction proceeds with a decline of free energy. When the equilibrium constant is less than 1.0 it is equivalent to saying that the reaction does not go far in the direction of completion. The standard free energy change is then positive and we must, therefore, put energy into the system to transform 1 mole of reactant into 1 mole of product under conditions where each is present at 1.0 molar concentration.

Let us now carry out a simple calculation of the equilibrium constant and the standard free energy change of a chemical reaction. We may use some data for the reaction

$$\text{glucose 1-phosphate} \rightleftharpoons \text{glucose 6-phosphate}$$

which is catalyzed by the enzyme *phosphoglucomutase*. This reaction occurs in the course of the breakdown of glycogen to lactate in muscle. If the reaction is started by adding the enzyme to 0.020 M glucose 1-phosphate solution at

TABLE 2–3. The Numerical Relationship Between the Equilibrium Constant and $\Delta G^{0\prime}$ at 25°C and pH 7.0

K'_{eq}	$\Delta G^{0\prime}$ kcal/mole
0.001	+4.09
0.01	+2.73
0.1	+1.36
1.0	0
10.0	−1.36
100.0	−2.73
1000.	−4.09

25°C and pH 7.0 it is found by chemical analysis of the medium that the reaction proceeds to an equilibrium at which the concentration of glucose 1-phosphate has decreased to 0.001 M, whereas the concentration of glucose 6-phosphate has risen from 0 to 0.019 M. The equilibrium constant is then

$$K_{eq} = \frac{[\text{glucose 6-phosphate}]}{[\text{glucose 1-phosphate}]} = \frac{0.019}{0.001} = 19$$

From the value $K'_{eq} = 19$, we may now calculate the standard free energy change of the phosphoglucomutase reaction

$$\Delta G^{0\prime} = -RT \ln K'_{eq}$$

$$= -1.987 \times 298 \times \ln 19$$

$$= -1.987 \times 298 \times 2.303 \times \log_{10} 19$$

$$= -1363 \times 1.28$$

$$\Delta G^{0\prime} = -1745 \text{ cal/mole}$$

Thus there is a decline in free energy of 1745 cal when 1.0 mole of glucose 1-phosphate is converted to 1.0 mole of glucose 6-phosphate at 25°C under conditions in which the concentration of each is maintained at 1.0 M. Ordinarily the standard free energy change is expressed in kilocalories (kcal) per mole. For the above reaction $\Delta G^{0\prime}$ is then -1.745 kcal.

Now we come to another important point. The presence of the enzyme (or any catalyst) in the system is irrelevant to the calculation; the enzyme merely accelerates the approach to equilibrium but does not influence the equilibrium point attained. Time and rate do not enter into our calculations, only the final state. Actually the standard free energy change we have calculated is the same whether the reaction is catalyzed by the enzyme or occurs by itself, however slowly. Furthermore, the intermediate mechanism or pathway taken by this chemical reaction is also irrelevant. When a reaction proceeds with a decline in free energy, we call it a *spontaneous* reaction. However, this term does not mean that the process will necessarily occur rapidly or "by itself." Thus, the combustion of glucose is a spontaneous process, but a pure, sterile solution of glucose left standing in oxygen at 20°C may not undergo oxidation for many years. If and when the reaction occurs, however, it will occur with a decline in free energy.

From the type of calculation outlined above, data have been assembled on the standard free energy changes of many chemical reactions that take place

TABLE 2–4. Standard Free Energy Changes at pH 7 and 25°C of Some Chemical Reactions

	$\Delta G^{0\prime}$ kcal/mole
Oxidation	
glucose + $6O_2 \rightarrow 6CO_2 + 6H_2O$	−686
lactic acid + $3O_2 \rightarrow 3CO_2 + 3H_2O$	−320
palmitic acid + $23O_2 \rightarrow 16CO_2 + 16H_2O$	−2,338
Hydrolysis	
sucrose + $H_2O \rightarrow$ glucose + fructose	−7.0
glucose 6-phosphate + $H_2O \rightarrow$ glucose + H_3PO_4	−3.3
glycylglycine + $H_2O \rightarrow$ 2 glycine	−2.2
Rearrangement	
glucose 1-phosphate $\rightarrow$ glucose 6-phosphate	−1.7
fructose 6-phosphate $\rightarrow$ glucose 6-phosphate	−0.4
Elimination	
malate $\rightarrow$ fumarate + H_2O	+0.75

in cells. Table 2–4 gives some representative values. It is seen that combustion reactions proceed with a relatively large decline in free energy, whereas hydrolysis reactions proceed with much smaller changes.

Reactions that have a negative standard free energy change are said to be *exergonic*; those that have a positive standard free energy change are said to be *endergonic*. It is convenient to think of exergonic reactions as "downhill" and endergonic reactions as "uphill."

2–8 THE FREE ENERGY CHANGE OF CHEMICAL REACTIONS UNDER NONSTANDARD CONDITIONS

The standard free energy change of a chemical reaction, such as those listed in Table 2–4, assumes that all reactants and products are present at the standard concentration of 1.0 M. But metabolites hardly ever exist in such high concentrations in cells, nor are they likely to be present in equimolar concentrations. Is it possible to calculate the free energy change of a chemical reaction at other than standard concentrations?

The answer is yes. For this purpose, we need to know not only the initial concentrations of the reactants and products but also the standard free energy change of the reaction. For the generalized reaction

$$aA + bB \rightleftharpoons cC + dD$$

the actual free energy change ΔG is given by the equation

$$\Delta G = \Delta G^{0\prime} + RT \ln \frac{[C]^c\,[D]^d}{[A]^a\,[B]^b}$$

in which $\Delta G^{0\prime}$ is the standard free energy change and $[A]$, $[B]$, $[C]$, and $[D]$ are the initial concentrations of reactants and products. R and T have their usual meaning and it is assumed that the temperature is 25°C (298°K) and the pH = 7.0.

With this equation we can now carry out calculation of the actual free energy change of a reaction in which the initial concentrations of the reactants and products are other than 1.0 M. From Table 2–1 we note that the hydrolysis of glucose 6-phosphate to yield glucose and phosphate is an exergonic reaction with a standard free energy change of −3.3 kcal. This reaction is catalyzed in the liver of vertebrates by the enzyme glucose 6-phosphatase and it is the primary reaction by which the free glucose of the blood is formed. Let us calculate the actual free energy change when this reaction is allowed to take place starting from initial concentrations of glucose 6-phosphate and glucose that are approximately physiological, and let it proceed to equilibrium. The concentration of glucose 6-phosphate in the liver cell is about 0.001 M and the concentration of glucose in the blood about 0.005 M. By substitution of the appropriate values in the above equation we obtain

$$\Delta G = -3300 + 1.98 \times 298 \times 2.303 \log_{10} \frac{[0.005]}{[0.001]}$$

$$= -3300 + 960$$

$$= -2340 \text{ cal}$$

$$= -2.34 \text{ kcal}$$

We see therefore that the free energy change for the hydrolysis of glucose 6-phosphate is substantially less under the conditions specified than the standard free energy change. From such calculations it is clear that the actual free energy changes of metabolic reactions under the conditions existing in the cell may be quite different from the standard free energy change. Nevertheless, for consistency we must use the arbitrarily defined standard free energy changes if we are to compare the energetics of chemical reactions quantitatively.

2–9 CATALYSIS AND THE ACTIVATION ENERGY

Because nearly all biochemical reactions are catalyzed by specific enzymes, it is important at this point to have some appreciation of how catalysts accelerate a chemical reaction without altering the final equilibrium reached. This is most easily understood with the aid of Fig. 2–3. When glucose 1-phosphate is placed in aqueous solution at pH 7.0, in the absence of the phosphoglucomutase enzyme, it will only very slowly undergo conversion to glucose 6-phosphate. Although this process is a "downhill" one, which ultimately proceeds with a free energy drop of 1.745 kcal, we see from the energy diagram that there is a barrier to this reaction, the *activation energy*. The energy content of molecules in a population follows a distribution curve, as is shown in Fig. 2–3. Only those molecules with a high energy content are likely to react to form the product. In order to make a reaction proceed faster we must raise the energy content of the entire population of molecules. One way of doing this is to heat the mixture; the heat absorbed by the molecules increases their internal energy and thus the likelihood of their colliding and reacting. However, there is another way in which the energy barrier can be overcome, namely, by adding a catalyst. In effect, the catalyst lowers the activation energy of the reaction by allowing a much larger fraction of the molecular population to react at any one time. The catalyst can do this because it can form an unstable intermediate complex or compound with the substrate, which quickly decomposes to the product and thus provides a "channel" or a "pass" through the activation energy barrier. Once this channel of low activation energy is found, the reaction proceeds rapidly, but because there is no difference whatsoever in the relative energy content of glucose 1-phosphate and glucose 6-phosphate in the uncatalyzed and catalyzed reactions, the point of equilibrium reached is the same in both cases.

2–10 OPEN AND CLOSED SYSTEMS

The thermodynamic principles just outlined apply to *closed* systems, which are defined as those that do not exchange matter with their surroundings. Furthermore, they apply only to systems that attain true thermodynamic equilibrium. However, living cells are *open* systems; they do exchange matter with their surroundings. Although the chemical components of living cells may appear to be stationary in concentration, they are not usually in true thermodynamic equilibrium but rather in a dynamic *steady state*, in which the rate of formation of a given component is exactly counterbalanced by an equal rate of removal or breakdown. There is one other important attribute of living organisms; they increase the entropy of the universe, as we shall see, and they are thus irreversible systems. Living cells, then, are open, irreversible systems which exist in a dynamic steady state.

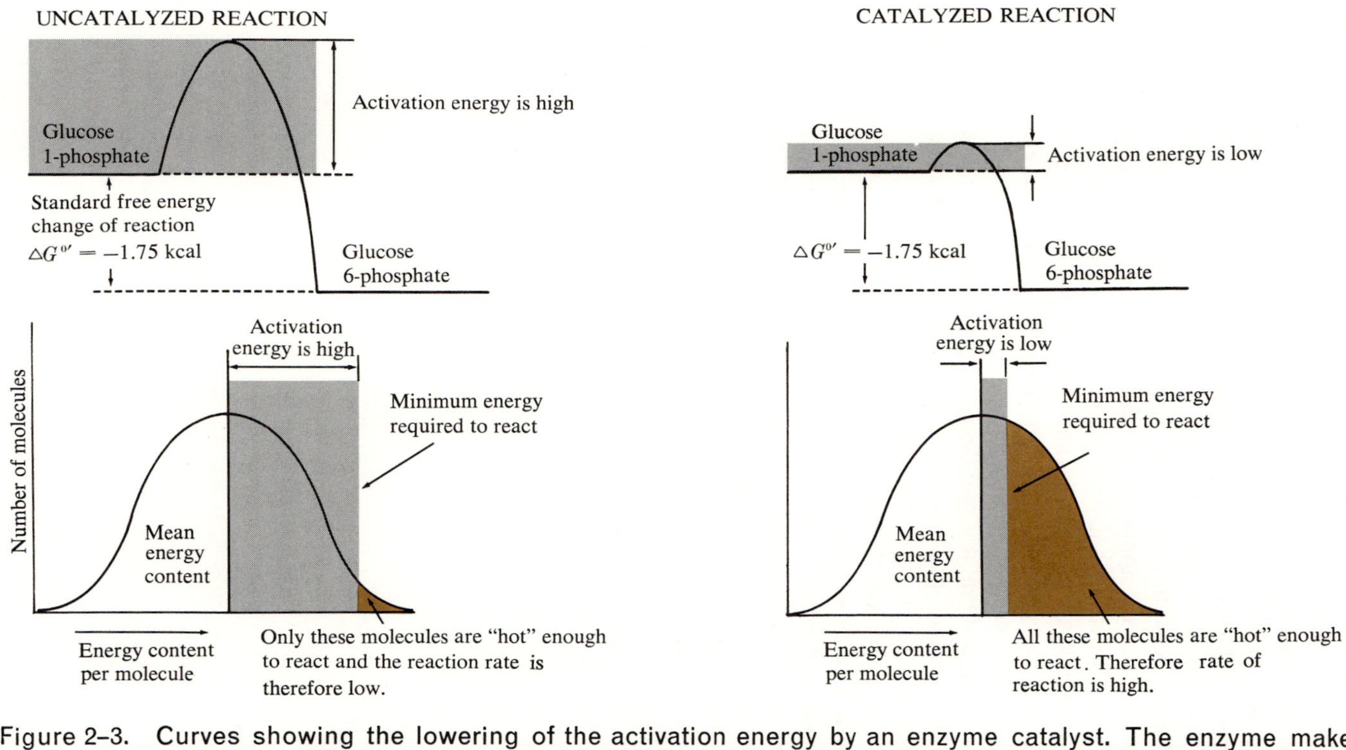

Figure 2–3. Curves showing the lowering of the activation energy by an enzyme catalyst. The enzyme makes possible a large increase in the fraction of substrate molecules having sufficient internal energy for the reaction to take place.

However, classical thermodynamic principles, some of which we have developed in this chapter, do not recognize the elements of time or rate; they deal only with the energy changes occurring as a system goes from an initial to an equilibrium state. Moreover, classical thermodynamics assumes idealized systems in which every change is frictionless. However, all *real* processes, the kind that take place in our everyday world, are accompanied by significant friction losses. Now there is a very important feature about friction that greatly complicates the application of equilibrium thermodynamics to many real processes. Friction is a time- or rate-dependent function, whereas in classical thermodynamics time does not enter into consideration. For example, the fraction of energy lost as friction in the bearings of an electric motor increases with the speed of the motor.

It might be argued that it is pointless to apply equilibrium thermodynamics to the analysis of biological processes, or, for that matter, to analysis of any real process. However, in biology, as well as in chemistry and physics, we must be able to analyze the simplest components of a complex system and describe exactly their structure and behavior under the simplest of conditions before we can understand how a series of components can interact together in a dynamic, continuing manner. Moreover, the laws of equilibrium thermodynamics comprise the starting point for the development of a new branch of thermodynamics able to cope with open systems. *Irreversible* or *nonequilibrium* thermodynamics is the name that has been given to this extension of thermodynamic theory. Although its formalism and mathematics are perhaps a little formidable for the elementary biology student, its application to biological systems is beginning to bring profound illumination.

3

THE ADENOSINE TRIPHOSPHATE SYSTEM AND THE TRANSFER OF CHEMICAL ENERGY

Adenosine triphosphate (ATP) was first isolated from acid extracts of muscle by Fiske and Subbarow in 1929, but a clear picture of its central role in energy transfer did not emerge until about 1940. Three sets of observations set the stage. First, it was found that ATP is generated by phosphorylation of ADP during the energy-yielding breakdown of glucose in extracts of animal tissues. Somewhat later, Engelhardt in the Soviet Union discovered that ATP is hydrolyzed to ADP by myosin, a major protein component of the contractile system of muscles. In still another line of investigation Cori and Cori found that ATP could promote the enzymatic formation of glycogen from glucose in the test tube, one of several observations that ATP may be important in biosynthesis of cell components.

In 1940 Lipmann assembled these and other items of evidence into a general hypothesis for the mechanism of energy transfers in the cell. He proposed that ATP is the agent that links energy-yielding and energy-requiring functions in the cell. Chemical energy yielded on degradation of fuel molecules is recovered by the coupled phosphorylation of ADP to yield ATP. The energy-rich ATP so formed then transfers its energy, by donation of its terminal "high-energy" phosphate group, to energy-requiring functions of the cell: biosynthetic processes, muscle contraction, and active transport against gradients.

In this chapter we shall examine the molecular structure of adenosine triphosphate, its biological distribution, and the specific features that endow it with the capacity to serve as the linking agent between energy-yielding and energy-requiring cellular reactions, the essence of its function in living cells.

3–1 STRUCTURE AND PROPERTIES OF ATP

Adenosine triphosphate occurs in all living cells–animal, plant, and microbial. It is present in concentrations of between 0.001 and 0.01 mole/liter of cell water, or about 0.5–5.0 mg/ml. By chromatographic methods, ATP is easily isolated from acid extracts of tissues. Its chemical structure (Fig. 3–1) was first deduced by Fiske and Subbarow in 1929, but the correctness of this structure was not established until nearly twenty years later, when ATP was first synthesized in the laboratory. ATP is a member of the family of compounds called *nucleotides*. Nucleotides contain a nitrogenous base, which may be either a purine or pyrimidine derivative, a 5-carbon sugar, and one or more phosphate groups. Thus ATP contains *adenine*, the 6-amino derivative of purine, the 5-carbon sugar D-*ribose*, which is attached to adenine through a glycosyl linkage, and a phosphate group in ester linkage at the 5-position of

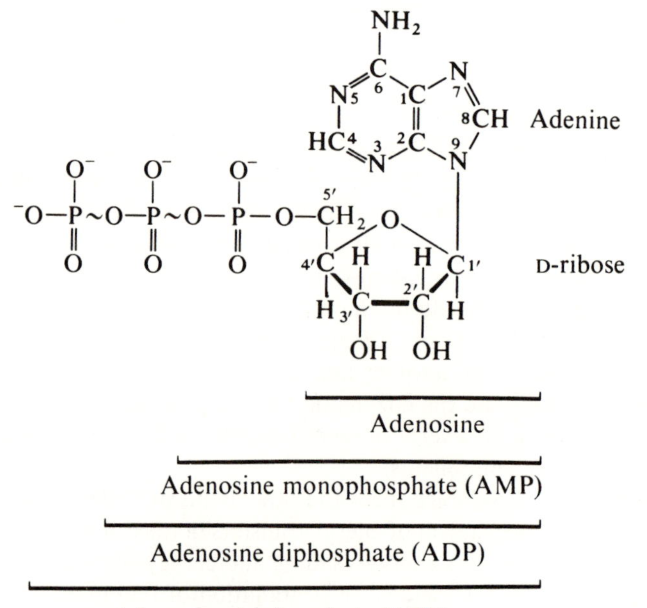

Figure 3–1. The structure of ATP, showing the ionized form at pH 7.0. The symbol ~ designates the "high-energy" bonds.

the ribose. When this compound, *adenosine monophosphate*, abbreviated AMP, contains a second phosphate group in anhydride linkage with the 5′-phosphate, we have *adenosine diphosphate* (ADP), and if there is a third phosphate group in linear linkage, we have *adenosine triphosphate* or ATP. ATP, ADP, and AMP occur in all cells.

The ATP molecule as it exists in the intact cell is highly charged; at pH 7.0 each of the three phosphate groups is nearly completely ionized. ATP therefore has close to four negative charges, which are concentrated around the linear polyphosphate structure. We will see later that this is an important feature of the ATP molecule with respect to its "high-energy" nature. ATP also forms stable complexes with certain divalent cations found in the cell, such as Mg^{2+} (Fig. 3–2). Actually most of the ATP in cells is present as the Mg^{2+} complex. This feature of the behavior of ATP is also related to its ability to act as an energy carrier.

3–2 THE STANDARD FREE ENERGY OF HYDROLYSIS OF ATP

When ATP was first isolated from muscle, it was suspected to have something to do with the energy of muscle contraction. Indeed it was later found that when ATP is incubated under appropriate conditions with muscle fibers, it undergoes enzymatic hydrolysis with the formation of ADP and inorganic phosphate

$$\text{ATP} + H_2O \longrightarrow \text{ADP} + \text{phosphate} \qquad (3\text{–}1)$$

It was also found rather early that this hydrolysis proceeds with a rather large liberation of heat when it is carried out in a calorimeter. But we are primarily interested in the standard free energy change of this reaction because it can give us the maximum amount of work this reaction can theoretically perform under conditions of constant temperature and pressure.

We have already seen that the standard free energy change of a chemical reaction can be determined from its equilibrium constant (Section 2–7). By measurement of the equilibrium constant of certain enzymatic reactions in

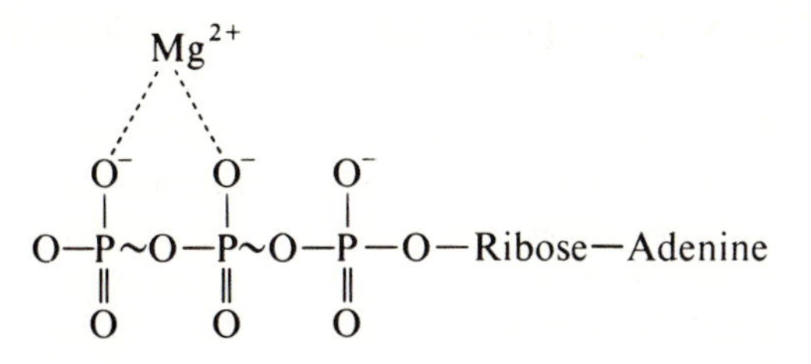

Figure 3–2. The Mg^{2+} complex of ATP.

which the terminal phosphate group of ATP is transferred, the standard free energy of hydrolysis of ATP has been calculated. For Eq. 3–1, in which only one phosphate group is hydrolyzed, with formation of ADP, the standard free energy of hydrolysis of ATP at pH 7.0 and 25°C is

$$\Delta G^{0\prime}_{\text{ATP}} = -7.3 \text{ kcal}$$

under conditions in which each reactant and product is present at 1.0 M concentration.

It is most unlikely that this value of −7.3 kcal represents the free energy of hydrolysis of ATP in the intact cell. One reason is that the concentrations of ATP, ADP, and phosphate in the cell are much lower than 1.0 M, the standard concentration defined by thermodynamic convention; moreover, these substances are not present in the cell in equimolar concentrations. An additional complication is the fact that intracellular Mg^{2+} forms complexes not only with ATP but also with ADP and phosphate, which differ in their affinity for Mg^{2+}. The presence of Mg^{2+} thus has the effect of shifting the equilibrium of ATP hydrolysis. If appropriate corrections are applied for all these factors, the free energy of hydrolysis of ATP to ADP and phosphate under intracellular conditions is believed to be nearer −12 kcal/mole. However, the free energy of hydrolysis of ATP in the cell is not necessarily constant; it may vary from one cell to another or it may vary from one time to another, depending on the intracellular pH, the Mg^{2+} concentration, and the existing concentrations of ATP, ADP, and phosphate. To be consistent in our energy calculations, we must of course employ only the *standard* free energy of hydrolysis, but it is clear from the considerations just outlined that rather large corrections must sometimes be applied to standard free energy data if they are to be precisely applicable to intracellular conditions.

The standard free energy of hydrolysis of ATP is significantly more negative than that of simple esters, glycosides, and amides, and of many phosphorylated compounds. For this reason, ATP has been called a *high-energy phosphate compound.* We will shortly see, however, that it is by no means unique in this respect.

3–3 WHY IS ATP A "HIGH-ENERGY" COMPOUND?

This question is tantamount to asking why the equilibrium of ATP hydrolysis lies further in the direction of completion than it does for simple esters or amides, or for other phosphate compounds, such as glycerol 3-phosphate. This follows from the quantitative relationship between the equilibrium constant and the standard free energy of hydrolysis developed in Section 2–7

$$\Delta G^{0\prime} = -RT \ln K_{\text{eq}}$$

which states that the larger the equilibrium constant, the more negative the standard free energy change (see Table 2–3).

There are two basic features of the ATP molecule that endow it with a relatively negative value for the standard free energy of hydrolysis; both are properties of its highly charged polyphosphate group. At pH 7.0 the linear polyphosphate portion of ATP has four negative charges that are very close to each other and that repel each other very strongly (Fig. 3–1). When the terminal phosphate bond is hydrolyzed, some of this electrostatic stress is relieved; the similar charges are separated as the ADP^{3-} and $phosphate^{2-}$ ions

$$ATP^{4-} + H_2O \longrightarrow ADP^{3-} + HPO_4{}^{2-}$$

Once the negatively charged products ADP and phosphate are separated, they will have very little tendency to approach each other again because their like charges repel each other. In contrast, phosphate esters of alcohols, for example, have no such repulsion forces operating between the products of their hydrolysis, because one of these, the alcohol, has no electrical charge

$$R—O—PO_3{}^{2-} + H_2O \longrightarrow R—OH + HO—PO_3{}^{2-} \qquad (3\text{–}2)$$

The second major factor contributing to the relatively negative value for the standard free energy of hydrolysis of ATP is the fact that the two products ADP and phosphate undergo stabilization as *resonance hybrids* as soon as they are formed. The electrons around the phosphorus and oxygen atoms of ATP and its hydrolysis products ADP and phosphate tend to seek that arrangement having the lowest possible energy. When ADP and phosphate are separated from each other, they have much less energy than they did when they were still covalently joined in the ATP molecule, because the cleavage of the terminal phosphate bond of ATP makes possible a new arrangement of electrons, one having a much lower energy content. Such resonance stabilization of the hydrolysis products is a major reason for the large negative value for the standard free energy of hydrolysis of ATP. By contrast, in the hydrolysis of a simple, low-energy phosphate ester of an alcohol, as in Eq. 3–2, the alcohol component formed as the hydrolysis product does not stabilize significantly after hydrolysis. We must always keep in mind that the standard free energy of hydrolysis of ATP or of any other compound is quite simply a measure of the *difference* in free energy of the initial reactants and the final products.

3–4 THE HIGH-ENERGY PHOSPHATE BOND

Those phosphorylated compounds having a strongly negative standard free energy of hydrolysis, such as ATP, are often spoken of as having *high-energy phosphate bonds*, and such bonds are universally designated by the symbol ~P. These terms are very useful to biochemists, but they may be a little misleading to the beginner. The term "high-energy phosphate bond" may imply that the energy spoken of is in the bond and that when the bond is split, energy is set free. This is not correct. In the ordinary usage of physical chemistry, bond energy is defined as the energy required to *break* a given bond between two atoms. Actually, relatively enormous energies are required to break chemical bonds, which would not exist if they were not stable. The term "phosphate bond energy" does *not* refer to the bond energy of the covalent linkage between the phosphorus atom and the oxygen or nitrogen atom; rather, it denotes the *difference* in energy content of the reactants and products. Phosphate bond energy is thus not localized in the chemical bond itself, but is a reflection of the relative energy content of the reactants and products.

3–5 THE STANDARD FREE ENERGY OF HYDROLYSIS OF OTHER PHOSPHATE COMPOUNDS

Biochemists for a long time have loosely classified the various phosphorylated compounds found in the cell into two groups, *high-energy phosphates* and *low-energy phosphates*. However, these terms have lost some of their meaning as more complete data on free energies of hydrolysis became available. Actually, there is no such sharp division. Furthermore, these terms obscure a very important and fundamental relationship among the various phosphorylated compounds in the cell, a relationship that is central to the whole idea of the transfer of chemical energy in cells.

Let us turn to Table 3–1 and examine the magnitude of the standard free

TABLE 3–1. Standard Free Energy of Hydrolysis of Phosphate Compounds

	$\Delta G^{0\prime}$ kcal/mole	Direction of Phosphate Group Transfer
Phosphoenolpyruvate	−14.8	↓
1,3-diphosphoglycerate	−11.8	
Phosphocreatine	−10.3	
Acetyl phosphate	−10.1	
ATP	−7.3	
Glucose 1-phosphate	−5.0	
Fructose 6-phosphate	−3.8	
Glucose 6-phosphate	−3.3	
3-phosphoglycerate	−2.4	
Glycerol 3-phosphate	−2.2	

energies of hydrolysis of various phosphate compounds. We see that these values have a wide range, and show no really sharp demarcation between high- and low-energy compounds. Moreover, the important point to note is that ATP, far from having the most negative value for $\Delta G^{0\prime}$ among the phosphate esters, actually has an intermediate value. The standard free energy of hydrolysis of ATP may be considered as forming the center or mid-point of a thermodynamic scale of phosphorylated compounds. This scale is arranged in order of the free energy of hydrolysis; those phosphate compounds high in the scale undergo more complete hydrolysis at equilibrium than those low in the scale. Put in another way, those high in the scale tend to lose their phosphate groups, and those lower in the scale tend to hold on to their phosphate groups. This thermodynamic scale is a quantitative measure of the affinity of each compound for its phosphoryl group.

Figure 3–3 shows the structure of some phosphate compounds important in energy transformations in the cell.

3–6 THE CENTRAL ROLE OF THE ATP-ADP SYSTEM

What, then, is so unique about ATP if it is only one of many high-energy compounds in the cell, some of which have a standard free energy of hydrolysis much more negative than that of ATP?

In the first place, ATP is unique because it occupies an *intermediate* position in the thermodynamic scale of phosphate compounds. It is the whole function of the ATP-ADP system to act as an intermediate linking system between phosphate compounds having a high phosphate group transfer potential, and other compounds having a low phosphate group transfer potential, by making possible the transfer of phosphate groups from the former to the latter. In brief, ADP serves as the specific enzymatic *acceptor* of phosphate groups from cellular phosphate compounds of very high potential. The latter are formed during the energy-yielding oxidation of foodstuffs in the cell and in them much of the energy of the foodstuff is conserved. The ATP so formed can now donate its terminal phosphate group enzymatically to certain specific phosphate acceptor molecules, such as glucose or glycerol, transforming them to their phosphate derivatives and thus raising their energy content. It is most important to note that this thermodynamic scale specifies the *direction* of enzymatic phosphate group transfer. If we start with equimolar concentrations of the phosphate donor and acceptor, phosphate groups will be transferred from compounds of high potential to acceptors of lower potential—*down* the scale (Fig. 3–4).

The second reason for the uniqueness of ATP in serving as a general energy carrier is that ATP and ADP are obligatory reactants in nearly all the enzymatic phosphate transfer reactions in the cell. One set of phosphate-transferring enzymes catalyzes transfer of phosphate groups from compounds of very

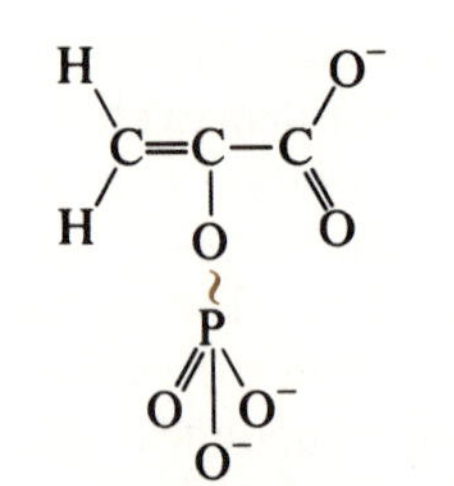

Phosphoenolpyruvate

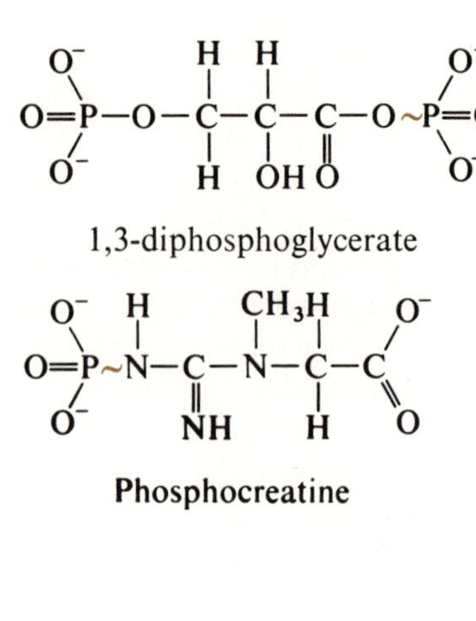

1,3-diphosphoglycerate

Phosphocreatine

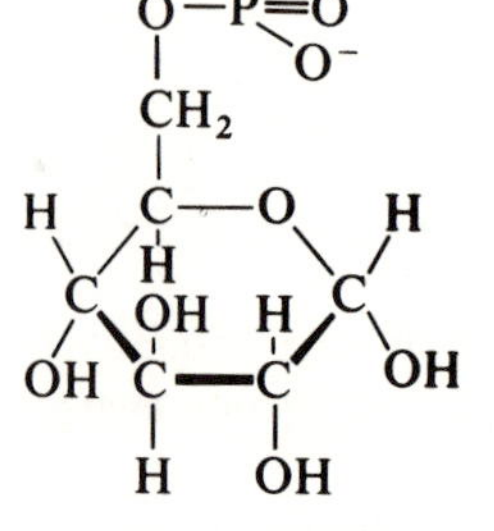

Glucose 6-phosphate

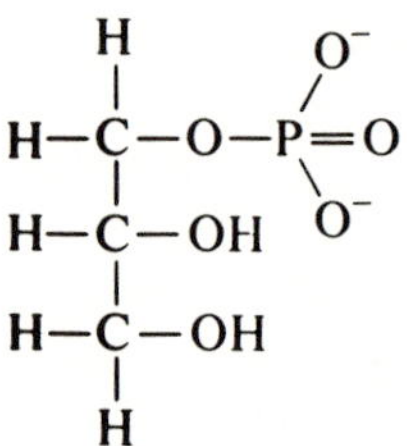

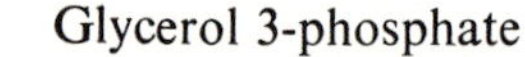

Glycerol 3-phosphate

Figure 3–3. The structure of some biologically important phosphorylated compounds.

high potential to ADP. A second set of enzymes catalyzes the reactions by which the terminal phosphate group of ATP is transferred to different low-energy phosphate acceptors. But there are *no* enzymes in cells that can transfer phosphate groups directly from high-energy donors to low-energy acceptors, without their being transferred via ATP. As we shall see (Chapters 4 and 5), "funneling" of all high-energy phosphate groups through the ATP system enables the cell to control the flow of energy in the simplest possible manner.

Figure 3–4 shows the principle involved in the action of the ATP-ADP system as a linking agent between high-energy phosphate donors, formed during oxidation of foods, and low-energy phosphate acceptors, which are energized by accepting a phosphate group, after which they can then do some form of cellular work. ATP and ADP thus constitute a "shuttle" for phosphate groups, the net flow of phosphate always being from high-energy to low-energy compounds. This diagram suggests that the terminal phosphate group of the ATP in the cell must undergo very rapid "turnover." This is

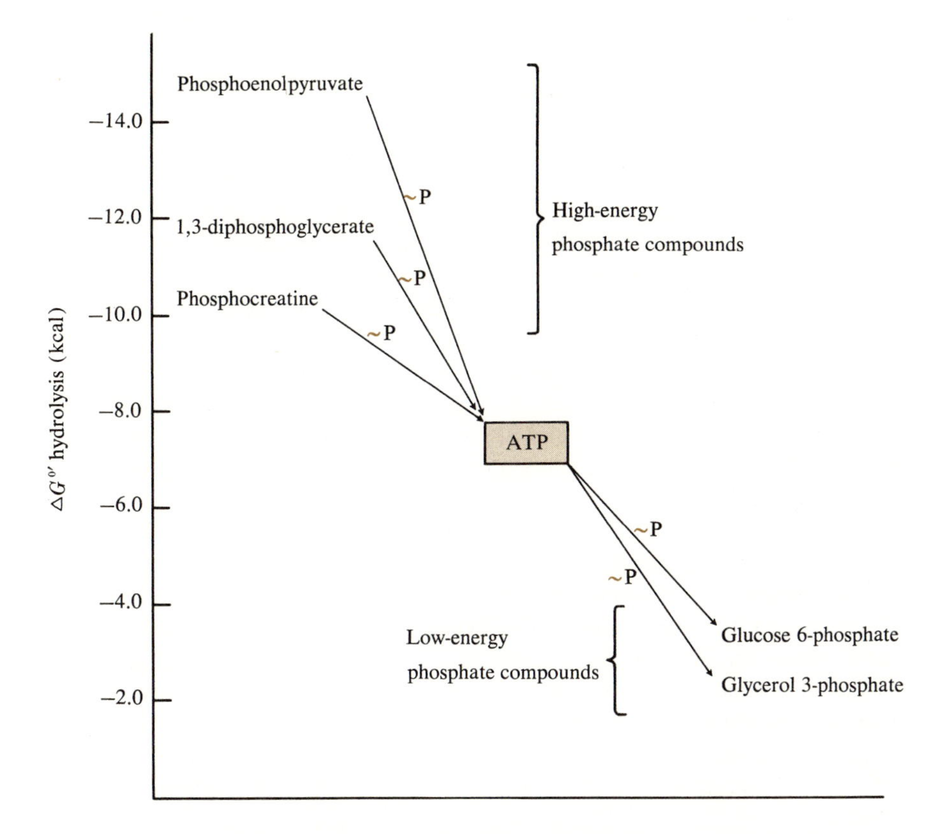

Figure 3–4. **The transfer of phosphate groups from high-energy phosphate donors to low-energy phosphate acceptors.**

in fact the case. In animal cells the half-time for turnover of the terminal phosphate groups of ATP has been estimated to be but a minute or two; in bacterial cells it is a matter of seconds.

3–7 THE PRINCIPLE OF THE COMMON INTERMEDIATE IN ENERGY-COUPLING

Adenosine triphosphate serves as a carrier of chemical energy between high-energy donors and low-energy phosphate acceptors because it is a *common intermediate* in both energy-delivering and in energy-requiring reactions of the cell. Let us clarify this term. When two chemical reactions occur sequentially, so that the product of the first reaction is the substrate or reactant of the second, we have a common intermediate. This is simply shown by the reactions

$$A + B \longrightarrow C + D \tag{3-3}$$

$$D + \mathrm{E} \longrightarrow F + G \tag{3-4}$$

In these two reactions, each of which is complete and can proceed independently of the other, the shaded component D is the common intermediate; it is a product of Eq. 3–3 and a reactant of Eq. 3–4.

The only way in which chemical energy can be transferred from one chemical reaction to another is for the two reactions to have a common intermediate which links them. Component D could thus be a carrier of chemical energy between Eq. 3–3 and Eq. 3–4; in contrast, component C could not function as such an energy carrier because it does not participate in Eq. 3–4. In brief, ATP can serve as a common intermediate in sequential reactions that are analogous to Eqs. 3–3 and 3–4.

The common intermediate principle is the basis of all biological energy transfers and we will see this principle operating over and over again. This is because nearly all the chemical processes in the cell proceed by *sequential* reactions in which the product of one reaction becomes the substrate of the next, and so on. ATP is by no means the only energy-carrying intermediate in each and every cellular reaction, but it is the common intermediate in the mainstream of energy transformations in the cell. Its role can be indicated in the following equations, in which X ~ P denotes a high-energy phosphate donor molecule and Y a phosphate acceptor; ATP is the common intermediate

$$\mathrm{X \sim P + ADP \longrightarrow X + ATP}$$

$$\mathrm{ATP + Y \longrightarrow ADP + Y{-}P}$$

3–8 CONSERVATION OF THE ENERGY OF OXIDATION AS ATP ENERGY

Now let us examine in some detail the enzymatic steps in the conservation of the energy of oxidation of foodstuffs in the form of a high-energy phosphate compound and the subsequent transfer of its phosphate group to ADP, recharging it to yield ATP. We will illustrate this process by an actual reaction occurring in the cell.

The oxidation, or dehydrogenation, of an aldehyde to a carboxylic acid in aqueous solution is known to proceed with a large decline in free energy, which varies with the nature of the aldehyde and the nature of the electron or hydrogen acceptor, but for simplicity we shall assume it is about the same as the standard free energy of hydrolysis of ATP

$$\underset{\underset{O}{\|}}{R\text{—}C\text{—}H} + H_2O \longrightarrow \underset{\underset{O}{\|}}{R\text{—}C\text{—}OH} + 2\,[H] \qquad (3\text{–}5)$$

$$\Delta G^{o\prime} = -7.3 \text{ kcal}$$

In the cell the oxidation of certain aldehydes may take place enzymatically in such a way that this energy is not simply lost but is in large part conserved in the form of a phosphate derivative of the oxidation product. For example, the aldehyde 3-*phosphoglyceraldehyde* is dehydrogenated to the acid 3-*phosphoglycerate* during the course of glucose degradation in the cell. However, this dehydrogenation is accompanied by the combination of one molecule of phosphate and one of ADP to form ATP. The overall equation for this reaction as it occurs in the cell is as follows, with RCHO symbolizing 3-phosphoglyceraldehyde, RCOOH 3-phosphoglycerate, and P_i free or inorganic phosphate

$$RCHO + P_i + ADP \longrightarrow RCOOH + ATP + 2[H] \qquad (3\text{–}6)$$

$$\Delta G^{0\prime} = 0.0 \text{ kcal}$$

We see in this equation that an aldehyde group (—CHO) was oxidized to a carboxylic acid (—COOH), a process which normally proceeds with liberation of a large amount of energy. We also see, however, that phosphate and ADP were simultaneously converted into ATP, a process that we already know requires input of energy and cannot occur spontaneously.

The mechanism by which this transfer of energy occurred is not really indicated by this overall reaction equation, but if we now break it down into its component reactions, the principle will become clear. The overall reaction (Eq. 3–6) occurs in two separate steps, each catalyzed by a separate enzyme. Let us now write these two reactions out in full, again allowing RCHO to represent 3-phosphoglyceraldehyde and RCOOH to represent 3-phosphoglyceric acid

$$R\text{—}CHO + P_i \longrightarrow \underset{\text{1,3-diphospho-glyceric acid}}{\underset{\underset{O}{\|}\quad\;\; \underset{O}{\|}}{R\text{—}C\text{—}O\text{—}\overset{\overset{OH}{|}}{P}\text{—}OH}} + 2[H] \qquad (3\text{–}7)$$

$$\mathrm{R{-}\underset{\underset{O}{\|}}{C}{-}O{-}\underset{\underset{O}{\|}}{\overset{\overset{OH}{|}}{P}}{-}OH} + \mathrm{ADP} \longrightarrow \mathrm{RCOOH} + \mathrm{ATP} \quad (3\text{–}8)$$

We see that the aldehyde is oxidized in Eq. 3–7, not directly to the carboxylic acid, but to a phosphorylated derivative of this acid, namely, 1,3-*diphosphoglyceric acid*; this is a mixed anhydride of phosphoric acid and a carboxylic acid (Fig. 3–3). This compound is remarkable for two things. In the first place, it is the common intermediate in the two sequential reactions (Eq. 3–7) and (Eq. 3–8); it is the product of reaction (Eq. 3–7) and a reactant of reaction (Eq. 3–8). The second important feature of this intermediate is that it has a standard free energy of hydrolysis substantially more negative than that of ATP (see Table 3–1). Thus, when the aldehyde RCHO was oxidized by the first enzyme, a large part of the free energy decline normally occurring when the aldehyde group is oxidized is now conserved in the form of the phosphate derivative of the carboxylic acid. In the second reaction, the carboxyl phosphate group of the 1,3-diphosphoglycerate is enzymatically transferred to ADP to form ATP. Thus the energy of oxidation of the aldehyde has been conserved in the form of ATP by two sequential reactions in which the high-energy phosphate derivative of a carboxylic acid was the common intermediate.

Now let us analyze quantitatively the energy transfers taking place in the pair of reactions described above to see exactly how effective they are in bringing about the conservation of the energy of oxidation as ATP energy. In Eq. 3–5 we saw that the oxidation of an aldehyde to a carboxylate group by removal of two H atoms proceeds with an approximate $\Delta G^{0\prime}$ of -7.3 kcal. In contrast, the energy-requiring process occurring during the coupled reactions (in Eqs. 3–7 and 3–8) was the formation of ATP from ADP and phosphate

$$P_i + \mathrm{ADP} \longrightarrow \mathrm{ATP} + H_2O$$

which must have required the input of at least 7.3 kcal/mole. We now see that the coupled synthesis of one molecule of ATP from ADP and phosphate conserved virtually all of the energy of the oxidation process in the form of the phosphate bond energy of the newly formed ATP molecule. The conservation of oxidative energy as ATP energy can take place only because the oxidation reaction and the phosphorylative reaction have a common intermediate.

This pair of reactions illustrates the working principle of the conversion of energy of oxidation of foodstuffs into the phosphate bond energy of ATP. Now let us see how the ATP so formed can in turn carry out chemical work in the cell.

3–9 UTILIZATION OF ATP ENERGY TO DO CHEMICAL WORK

As an example of the chemical reaction pattern by which the phosphate bond energy of ATP is utilized to do chemical work, we shall consider the mechanism by which cells form the amide bond of glutamine. Glutamine is one of

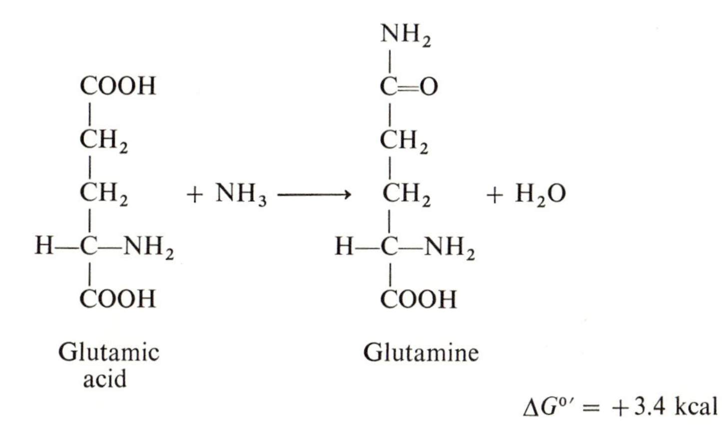

$\Delta G^{0\prime} = +3.4$ kcal

Figure 3–5. The formation of glutamine from glutamic acid and ammonia.

the amino acid building blocks of proteins, and its formation from glutamic acid and ammonia is one of the many energy-requiring biosynthetic reactions in living cells. The formation of glutamine from glutamic acid and ammonia is an endergonic reaction since it has a positive standard free energy change (Fig. 3–5). Because of this large positive value of $\Delta G^{0\prime}$, the point of equilibrium is far to the left and it is clear that very little glutamine can be made by this reaction at any reasonable concentrations of its precursors.

In the living cell the synthesis of glutamine is accomplished by a different pathway, in which formation of glutamine is coupled to the breakdown of ATP to ADP and phosphate. This occurs through two sequential reactions catalyzed by the enzyme *glutamine synthetase*

$$\text{ATP} + \text{glutamic acid} \longrightarrow \text{glutamyl phosphate} + \text{ADP} \quad (3\text{–}9)$$

$$\text{glutamyl phosphate} + NH_3 \longrightarrow \text{glutamine} + P_i \quad (3\text{-}10)$$

$$\text{Sum: ATP} + \text{glutamic acid} + NH_3 \longrightarrow \text{glutamine} + \text{ADP} + P_i$$

$\Delta G^{0\prime} = -3.9$ kcal

We see that these two reactions have a common intermediate, glutamyl phosphate, a phosphate derivative of glutamic acid (Fig. 3–6). This

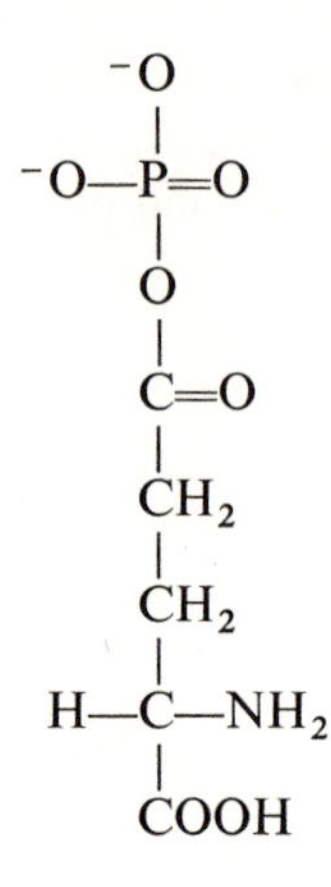

Figure 3–6. Glutamyl γ-phosphate.

compound is a product of the first reaction and is used up in the second; both its formation and utilization occur on the active or catalytic site of the enzyme molecule. In the first reaction, the terminal phosphate group of ATP was transferred to glutamic acid, and with it some of the energy of ATP. In the second reaction, the glutamyl phosphate exchanges its phosphate group for ammonia to form the amide linkage. The overall reaction, as is seen from the negative sign of the $\Delta G^{0\prime}$ value, is exergonic and tends to proceed in the direction of glutamine formation.

If we now dissect this sequence of reactions to analyze the energy exchanges, we can see that the energy-yielding process is the hydrolysis of ATP and the energy-requiring process is the formation of glutamine:

Energy-yielding process

$$\text{ATP} + H_2O \longrightarrow \text{ADP} + P_i$$

$$\Delta G^{0\prime} = -7.3 \text{ kcal}$$

Energy-requiring process

$$\text{glutamic acid} + NH_3 \longrightarrow \text{glutamine} + H_2O$$

$$\Delta G^{0\prime} = +3.4 \text{ kcal}$$

These processes do not take place independently as written but in the form of the coupled Eqs. 3–9 and 3–10, which together have a net decline in free energy of −3.9 kcal. Thus the amide bond of a molecule of glutamine is synthesized from glutamic acid and ammonia at the expense of a high-energy phosphate bond of a molecule of ATP.

This is the basic reaction pattern utilized by the cell in making many different kinds of chemical bonds, such as ester linkages, peptide bonds, and glycosidic linkages, which otherwise would not form in any quantity in dilute aqueous systems. High-energy phosphate bonds of ATP are sacrificed to make such chemical bonds through enzyme-catalyzed reactions in which the breakdown of ATP shares a common intermediate with the synthesis of the new linkage. In the simple example we have discussed, a phosphate group is transferred directly from ATP to the compound undergoing activation. However, we shall see later (Chapters 7 and 8) that in some biosynthetic reactions chemical groups other than phosphate are employed to activate the building-block molecules, but the cleavage of high-energy phosphate bonds of ATP is always the process that ultimately provides the necessary chemical energy.

3–10 THE LINKAGE OF AN ENERGY-YIELDING OXIDATION TO AN ENERGY-REQUIRING BIOSYNTHETIC REACTION

Now let us put together the two energy-transfer processes we have been discussing, namely the conservation of the energy of oxidation of an aldehyde as the phosphate bond energy of ATP, and the utilization of the ATP energy to carry out the formation of the amide linkage of glutamine

$$\mathrm{RCHO} + \mathrm{P_i} \longrightarrow 2[\mathrm{H}] + \mathrm{RCOOPO_3H_2} \qquad (3\text{–}7)$$

$$\mathrm{RCOOPO_3H_2} + \mathrm{ADP} \longrightarrow \mathrm{RCOOH} + \mathrm{ATP} \qquad (3\text{-}8)$$

$$\mathrm{ATP} + \text{glutamic acid} \longrightarrow \mathrm{ADP} + \text{glutamyl phosphate} \qquad (3\text{–}9)$$

$$\text{glutamyl phosphate} + \mathrm{NH_3} \longrightarrow \text{glutamine} + \mathrm{P_i} \qquad (3\text{–}10)$$

$$\text{Sum: } \mathrm{RCHO} + \text{glutamic acid} + \mathrm{NH_3} \longrightarrow 2[\mathrm{H}] + \mathrm{RCOOH} + \text{glutamine} \qquad (3\text{–}11)$$

$$\Delta G^{0\prime} = -3.9 \text{ kcal}$$

The overall equation (Eq. 3–11), the sum of the individual reactions (Eqs. 3–7, 3–8, 3–9, 3–10), shows that the energy yielded by oxidation of the aldehyde to an acid was used to form glutamine from glutamic acid and ammonia. This energy transfer did not take place directly but required a series of four sequential reactions, coupled at each step by a common intermediate, which in each case is shown in shaded boxes. We clearly see from this sequence that ATP constitutes the common intermediate linking the energy-conserving and the energy-requiring reactions. The overall reaction proceeds in the direction of completion because the overall $\Delta G^{0\prime}$ is negative. Moreover, we also note that the overall process is quite efficient, since it has brought about the formation of a bond requiring 3.4 kcal at the expense of an oxidation yielding about 7.0 kcal, or an efficiency of $3.4/7.0 \times 100 = 50$ per cent.

This simple example illustrates the general working principle by which ATP serves as the energy-carrier between the energy-yielding degradation of fuel molecules and the energy-requiring synthesis of cell components from simpler precursors. In the following chapters we shall continue to enlarge on this plan.

4

GENERATION OF ATP IN ANAEROBIC CELLS

In this chapter we shall consider the relatively simple enzymatic mechanisms by which cells extract useful energy from foodstuff molecules in the complete absence of oxygen. We will recall that in all types of cells the endergonic formation of ATP from ADP and phosphate is coupled or linked to oxidation reactions, which generally proceed with a large decline in free energy. In aerobic cells, oxygen is the final electron acceptor or oxidant of the fuel. However, when cells live in the absence of oxygen, the fuel molecule is broken down into two or more fragments; one of these fragments then becomes oxidized by another. Coupled to this oxidation is the formation of ATP. Now let us see how this process is carried out by the multienzyme system catalyzing the anaerobic breakdown of glucose to lactic acid.

4–1 THE BIOLOGICAL ROLE OF ANAEROBIC FERMENTATION

In the entire realm of living organisms, only a few species are strictly anaerobic—that is, can live only in the absence of oxygen. Most are microorganisms, particularly those species that live in surroundings having little or no oxygen, in soils, in deep waters, or in marine mud. Among these are some *pathogens* (disease-causing organisms). An example is the soil bacterium *Clostridium welchii*, the cause of gas gangrene in wound infections. Since there are only relatively few completely anaerobic organisms, we might think it is not worth a chapter of this book to consider how these forms of life extract energy from glucose and other foodstuff molecules.

As it happens, however, there are a great many organisms that can live either in the absence or presence of oxygen. These include not only a large number of microorganisms, but also many higher animals and plants. Even certain tissues of the human organism are able to function either aerobically or anaerobically. Cells that can live under either aerobic or anaerobic conditions are called *facultative* cells. The significant point is that when oxygen is absent from their environment they are able to extract energy from glucose by the same type of anaerobic fermentation mechanisms that are used by the strict anaerobes, whereas in the presence of oxygen, they prefer to oxidize their foodstuffs completely with oxygen.

It now appears probable that in the course of biological evolution the first living organisms were simple anaerobic cells able to live by fermentation mechanisms only. Ultimately, when oxygen first appeared in the atmosphere as a product of photosynthesis, some cells acquired the capacity to use oxygen as an oxidant. In facultative organisms the capacity to live without oxygen has been retained, allowing them great metabolic flexibility. Furthermore, in most facultative cells the anaerobic or fermentative means of extracting energy from glucose is a preparatory step, which is followed by a second stage in which the fermentation products are oxidized by molecular oxygen. These relationships are shown schematically in Fig. 4–1. Under anaerobic conditions glucose is broken down to lactic acid, at which point fermentation stops. When oxygen is present, lactic acid does not accumulate; rather the products of fermentation are oxidized further to CO_2 and H_2O. Thus the fermentation reactions are common to both the aerobic and anaerobic breakdown of sugar in many facultative organisms.

Various microorganisms differ from each other in the end products they produce from glucose. For example, when yeast cells live anaerobically, they ferment glucose to ethyl alcohol and CO_2. On the other hand some bacteria ferment glucose to acetone, some to butanol, some to acetic acid and ethanol, and still others to lactic acid. The characteristic end products of glucose fermentation are commonly used to identify or classify different microorganisms. In all fermentations of glucose in anaerobic organisms the glucose molecule is always broken down in such a way as to yield two or more fragments, one of which is oxidized by another. Some of the energy yielded in this oxidation-reduction process is conserved as ATP energy.

4–2 GLYCOLYSIS

Let us now consider the mechanism of breakdown of glucose into two molecules of lactic acid, the type of fermentation known as *glycolysis*, which means

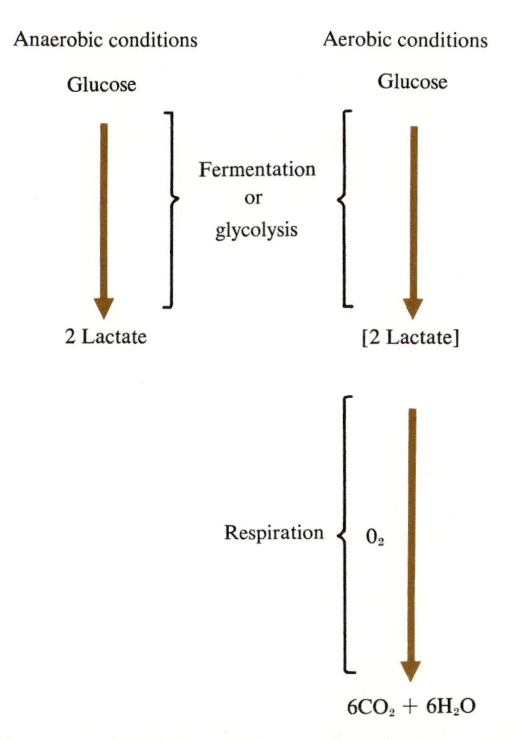

Figure 4–1. Glucose breakdown in a facultative organism under anaerobic and aerobic conditions.

"dissolution of sugar." Glycolysis is the most widespread and best understood of the anaerobic mechanisms for utilizing glucose among different forms of life; it is also the type of glucose breakdown that occurs in the cells of higher animals.

The anaerobic breakdown of glucose into two molecules of the 3-carbon compound lactic acid may be written as follows

$$\underset{\text{glucose}}{C_6H_{12}O_6} \longrightarrow \underset{\text{lactic acid}}{2CH_3\text{—}\overset{\overset{\displaystyle H}{|}}{\underset{\underset{\displaystyle OH}{|}}{C}}\text{—}COOH} \qquad (4\text{–}1)$$

$$\Delta G^{o\prime} = -47 \text{ kcal}$$

We note that no oxygen is required, nor is there any net oxidation or reduction, since the ratio carbon : hydrogen : oxygen of the reaction product is identical to that of the glucose, namely, 1 : 2 : 1. This reaction occurs with a rather large decline in free energy, and thus it will proceed spontaneously, in the thermodynamic sense, providing there is a mechanism or pathway available to catalyze it.

However, the anaerobic breakdown of the glucose molecule to lactate in living cells is incompletely described by Eq. 4–1. Rather it proceeds by a mechanism in which phosphate, ADP, and ATP also participate, in the following overall equation

$$\text{glucose} + 2\text{P}_\text{i} + 2\text{ADP} \longrightarrow 2\ \text{lactate} + 2\text{ATP} + 2\text{H}_2\text{O} \qquad (4\text{–}2)$$

$$\Delta G^{0\prime} = -32.4\ \text{kcal}$$

This equation may actually be considered as the sum of two linked processes. In one, glucose is broken down to two molecules of lactate, and in the other, two molecules of ATP are formed from two of ADP and two of phosphate. These two processes do not take place independently; rather they are coupled so that glucose breakdown cannot occur without ATP formation and ATP formation cannot occur without glucose breakdown.

We are already in a position to draw a conclusion about the energetics of anaerobic glucose breakdown in the cell. Note that Eq. 4–2 proceeds with a free energy decline that is 14.6 kcal less than that of Eq. 4–1. From thermodynamic principles we know this must be so because at least 7.3 kcal are required to generate each mole of ATP from ADP and phosphate. We may thus conclude that in the intact cell a significantly large part of the free energy loss that occurs when glucose is broken down to lactate is conserved in the form of ATP; in fact, some (14.6/47)100 or about 31 per cent. The conserved energy is the cell's profit from the fermentation of glucose. Cells do not break down glucose just to dispose of it; they carry out this process largely to generate ATP from ADP and phosphate.

Equation 4–2 represents the principle of all anaerobic fermentations of foodstuffs by anaerobic organisms; the breakdown of the foodstuff molecule always proceeds with a stoichiometrically coupled formation of ATP from ADP and phosphate. The ATP so formed conserves a large fraction of the energy of the original foodstuff molecule.

4–3 ENZYMES

It is in order at this point to review very briefly the molecular nature of enzymes and their action. An enzyme is a specialized protein molecule that has the capacity to accelerate the rate of a chemical reaction. Enzymes are true catalysts; they do not influence the point of equilibrium of the reaction they catalyze, nor are they used up during catalysis. Like other catalysts, enzymes lower the activation energy of the reaction they catalyze (Section 2–8). Over

1000 enzymes are now known, each of them capable of catalyzing a specific chemical reaction. About 150 enzymes have been isolated in pure crystalline form.

The amount of an enzyme in a tissue or cell extract can be measured by the catalytic effect it produces (Fig. 4–2). Under defined conditions of temperature, pH, and the concentration of the substance on which it acts, which is called the *substrate*, the rate of an enzyme-catalyzed reaction is proportional to the concentration of the enzyme. An enzyme becomes "saturated" with its substrate if the latter is present at a high enough concentration (Fig. 4–2) and under these conditions it shows its maximum velocity of catalysis. The saturation phenomenon has led to the hypothesis that an enzyme E combines transiently with its substrate S to yield an enzyme-substrate complex ES whose rate of breakdown to yield free enzyme and the product P may be rate limiting

$$\mathrm{E + S \longrightarrow ES}$$

$$\mathrm{ES \longrightarrow E + P}$$

Because enzymes are very specific in their action it has been postulated that there is a lock-and-key fit of the substrate molecule to a small patch on the surface of the very large enzyme molecule. This patch is called the *active site* or *catalytic site*. Because of the specific geometrical relationship of the chemical groups that combine with the substrate, the active site can accept only molecules having a complementary fit (Fig. 4–2).

Recent research has shown that all enzymes consist of one or more polypeptide chains that are tightly folded into a compact globular structure. In the formation of the enzyme-substrate complex the enzyme molecule is deformed slightly, to yield what has been called an "induced fit" of the enzyme to the substrate. This change places some strain on the geometry of the substrate molecule, rendering it more susceptible to attack by specific catalytic groups of the enzyme. Once the substrate has been acted upon, the enzyme molecule returns to its native shape, which causes it to release the products of the reaction. The enzyme molecule then combines with a second substrate molecule and repeats the cycle. In one minute a single enzyme molecule may carry out as many as several million such catalytic cycles.

Many enzymes can be inhibited by specific poisons that may be structurally related to their normal substrate; such inhibitors are very useful in analyzing enzyme-catalyzed reactions in cells and tissues, as we shall see.

Enzyme molecules are the instruments for all the energy conversions in the cell, and we will find that sometimes they are very elaborately adapted for this function. When enzymes act in a sequence so that the product of one enzyme becomes the substrate for the next, and so on, we have a multienzyme system.

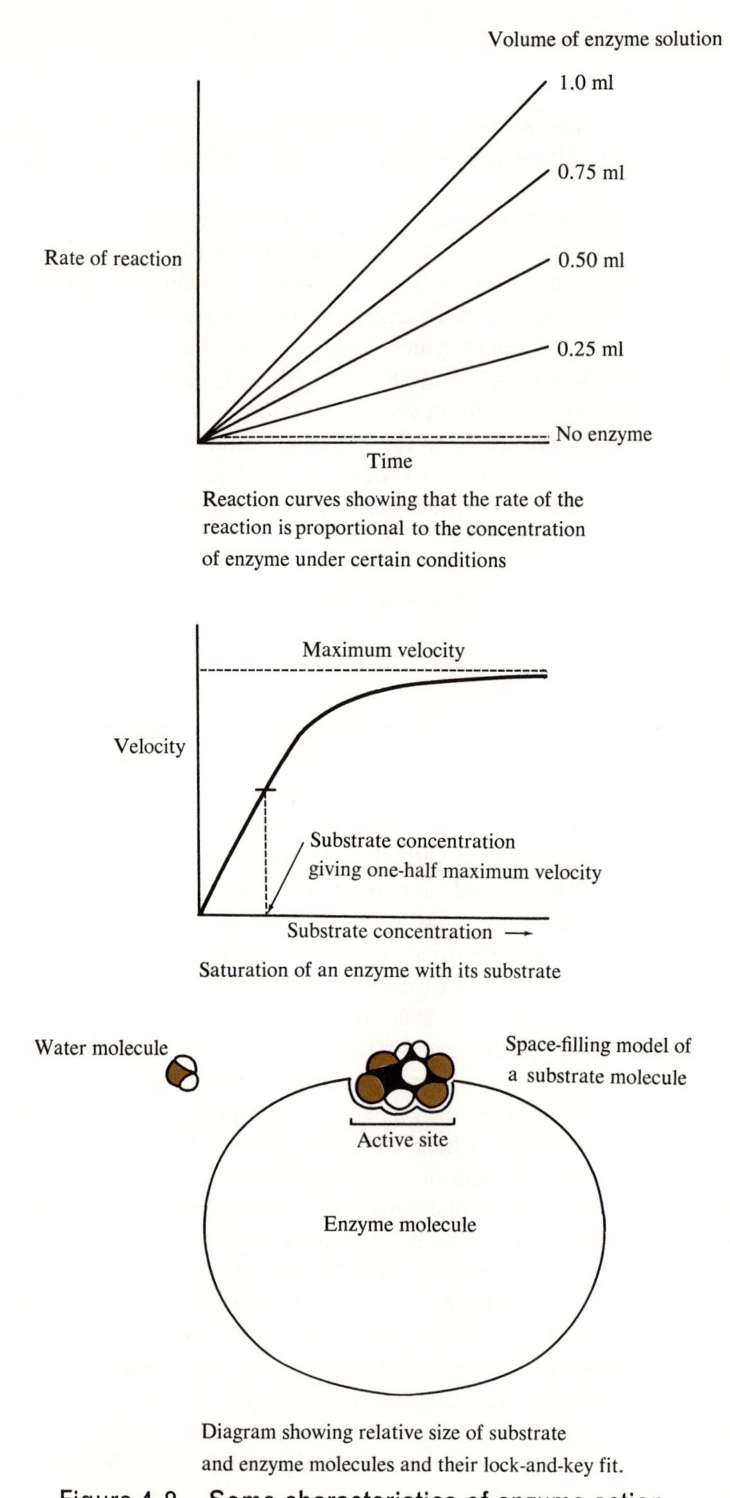

Reaction curves showing that the rate of the reaction is proportional to the concentration of enzyme under certain conditions

Saturation of an enzyme with its substrate

Diagram showing relative size of substrate and enzyme molecules and their lock-and-key fit.

Figure 4–2. Some characteristics of enzyme action.

Certain enzymes are responsible for the regulation of the rate of multi-enzyme sequences, for which they function as "pacemakers." Such enzymes are called *regulatory* or *allosteric* enzymes. We shall see later how the rate of glycolysis is controlled by a regulatory enzyme.

4–4 THE SEQUENTIAL STEPS IN GLYCOLYSIS

The modern era of research on the mechanism of anaerobic fermentations began in 1897 with the discovery of Buchner that cell-free extracts of yeast can catalyze alcoholic fermentation. Some years later Harden and Young obtained the first evidence that phosphate is required in this process and that a phosphorylated sugar, namely fructose 1,6-diphosphate, accumulates in such extracts under certain conditions. When this compound was added to a fresh yeast extract it was found to be fermented to ethanol and CO_2, showing that it is an intermediate. It was not long before glucose 6-phosphate and fructose 6-phosphate were isolated from fermenting yeast extracts. These early observations strongly suggested that fermentation consists of a number of consecutive enzyme-catalyzed reactions in which phosphorylated sugars are intermediates. In the 1920's it was found that extracts of skeletal muscle catalyzed formation of lactic acid from glucose; the same phosphorylated sugars acting in alcoholic fermentation were also found to be active in muscle.

Another important experimental approach came from the addition of inhibitory poisons. For example the compound *iodoacetic acid* (ICH_2COOH) inhibits glycolysis in muscle extracts and causes the 3-carbon sugars glyceraldehyde 3-phosphate and dihydroxyacetone phosphate to accumulate. Sodium fluoride also inhibits glycolysis and in its presence 3-phosphoglyceric and 2-phosphoglyceric acid accumulate. Today we know that these substances inhibit specific enzymes in the glycolytic pathway. By such methods the entire sequence of reactions leading from glucose to lactic acid in muscle or to ethanol and CO_2 in yeast were ultimately elucidated. The sequence of reactions leading from glucose to pyruvic acid is often called the Embden-Meyerhof pathway, after the two German biochemists who postulated and experimentally analyzed critical steps in this pathway in the late 1920's and early 1930's. Today we know that glycolysis requires a total of eleven specific enzyme molecules, acting in sequence in such a manner that the product of the first enzyme-catalyzed reaction becomes the substrate or reactant of the next. There are thus eleven distinct chemical reactions in the overall breakdown of glucose to lactate. Each of the eleven enzymes catalyzing these reactions has been isolated in substantially pure form and most of them have been crystallized. With these highly purified enzymes it is possible to reconstruct portions of the glycolysis sequence in the test tube. By such reconstruction approaches, the mechanism and dynamics of glycolysis can be scrutinized closely.

In Fig. 4–3 are shown the sequential chemical changes through which the

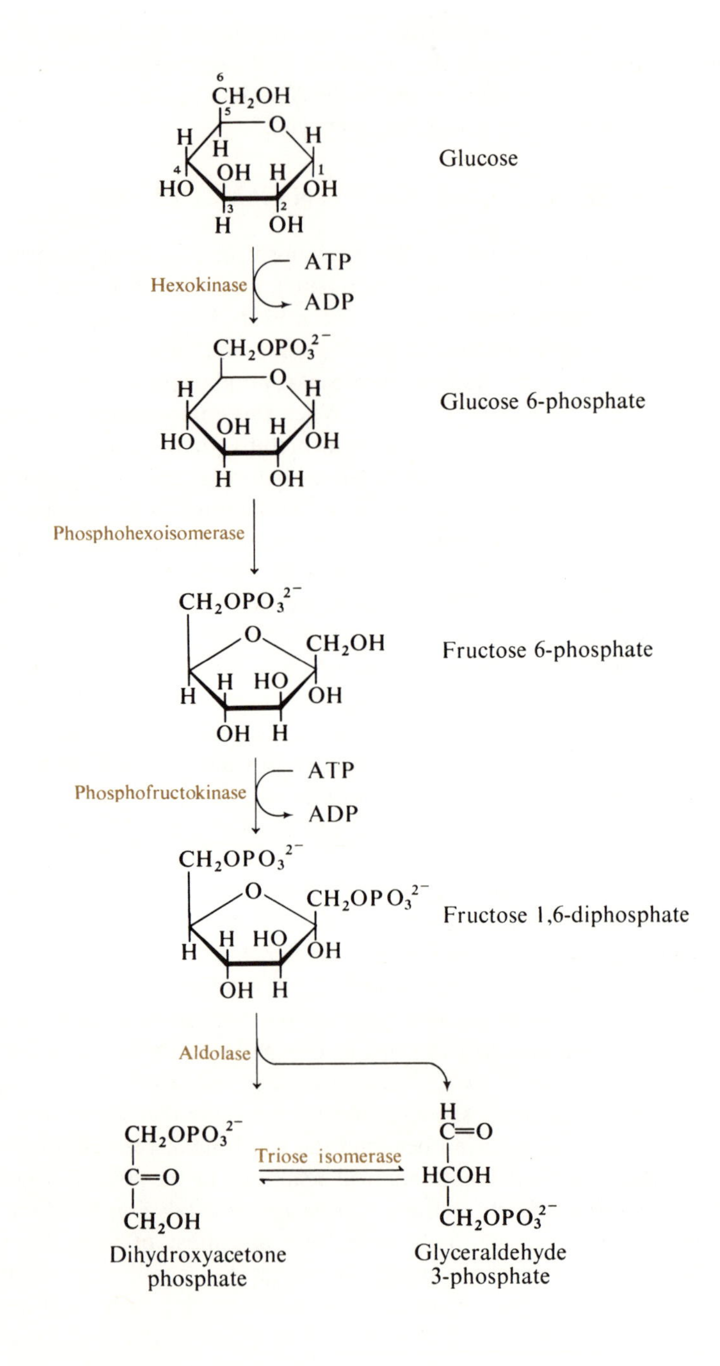
Glucose
Hexokinase
ATP
ADP
Glucose 6-phosphate
Phosphohexoisomerase
Fructose 6-phosphate
Phosphofructokinase
ATP
ADP
Fructose 1,6-diphosphate
Aldolase
Triose isomerase
Dihydroxyacetone phosphate
Glyceraldehyde 3-phosphate

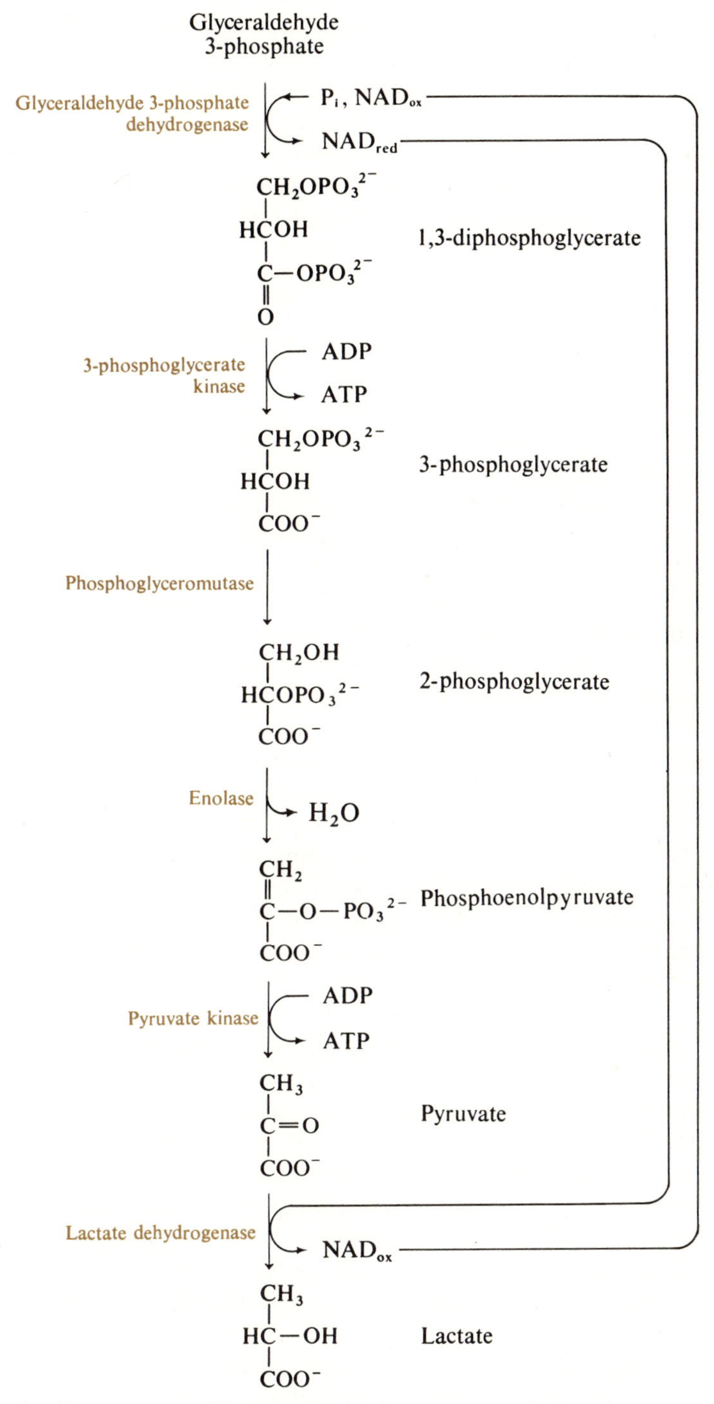

Figure 4–3. The sequential reactions of glycolysis.

glucose molecule is broken down to lactate. The scheme may look rather complex at first, but we shall find it has a basic simplicity in its overall features. In the "flowsheet" of glycolysis there are three distinct flows or pathways, which we shall examine in turn: (1) the pathway of the carbon atoms of glucose, (2) the pathway of electrons, and (3) the pathway of phosphate groups.

4–5 THE PATHWAY OF CARBON IN GLYCOLYSIS

Now let us look more closely at the pattern of the glycolytic reactions (Fig. 4–3) to see if we can discern the molecular principles and devices by which this sequence of reactions makes possible conservation of some of the energy of glucose as ATP energy.

First let us examine the pathway of degradation of the carbon atoms of glucose. Glucose, after two priming phosphorylation steps, is converted into fructose 1,6-diphosphate, which is then enzymatically split into two different 3-carbon sugars, each in phosphorylated form, namely, 3-phosphoglyceraldehyde and dihydroxyacetone phosphate. These are reversibly interconvertible by the action of the enzyme *triose phosphate isomerase*, which catalyzes an aldose-ketose transformation. From this point on we deal with the two molecules of the 3-carbon sugar 3-phosphoglyceraldehyde, each of which is ultimately transformed into the 3-carbon compound lactic acid.

The pathway of the carbon atoms of glucose is shown in Fig. 4–4. We note that carbon atoms 3 and 4 of the starting glucose ultimately become the carboxyl atoms of lactic acid, whereas carbon atoms 1 and 6 of the glucose become the methyl carbon atoms of lactic acid. That the glucose molecule actually cleaves in this manner in intact cells and tissues has been repeatedly confirmed by use of glucose labeled in different carbon atoms with an isotope of carbon, such as ^{13}C or ^{14}C.

4–6 THE PATHWAY OF ELECTRONS IN GLYCOLYSIS

Earlier we pointed out that in anaerobic fermentation one of the fragments of glucose becomes oxidized by the other. Let us now analyze the glycolytic sequence to see just where this oxido-reduction occurs.

In fact, we have already examined one aspect of it in some detail in Section 3–8 namely, the oxidation of 3-phosphoglyceraldehyde to 1,3-diphosphoglycerate. When we considered this oxidation before, we did not specify the nature of the oxidant but simply said that two hydrogen atoms, or their equivalent, two electrons, were removed from the aldehyde by an enzyme. Now, however, we can identify the ultimate oxidant or electron acceptor for the two electrons removed from 3-phosphoglyceraldehyde. It is another

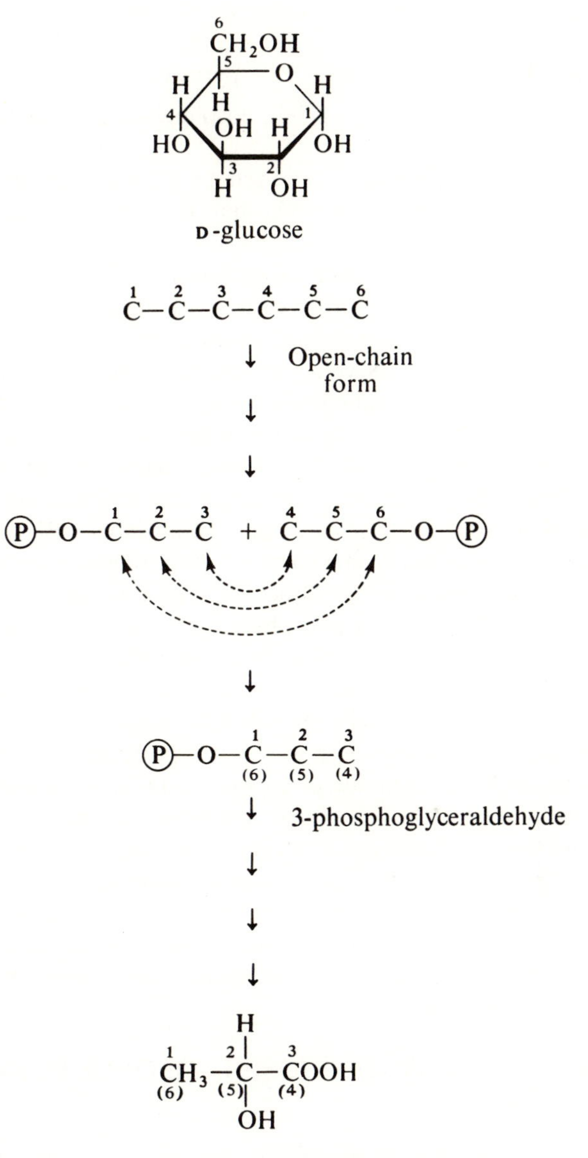

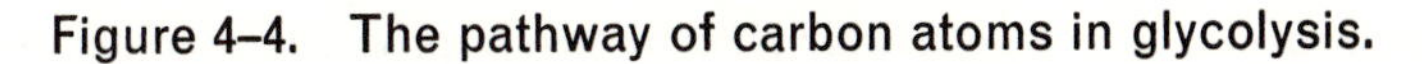
Figure 4–4. The pathway of carbon atoms in glycolysis.

3-carbon intermediate in the glycolytic sequence, namely pyruvate, which accepts these electrons and becomes reduced to lactate

$$\text{3-phosphoglyceraldehyde} + P_i + \text{pyruvate} \longrightarrow \text{1,3-diphosphoglycerate} + \text{lactate} \qquad (4\text{–}3)$$

Thus 3-phosphoglyceraldehyde is oxidized to 1,3-diphosphoglycerate, and pyruvate is reduced to lactate; one 3-carbon fragment of glucose has been oxidized by another.

But what is the role of the substance cryptically referred to in Fig. 4–3 as NAD? This compound is an electron carrier; by the ubiquitous principle of the common intermediate, it can carry electrons from one reaction to another. NAD can exist in oxidized (NAD_{ox}) and reduced (NAD_{red}) forms: NAD_{ox} accepts electrons from 3-phosphoglyceraldehyde and thus becomes reduced to NAD_{red}, which then donates electrons to pyruvate, regenerating NAD_{ox}.

NAD is the abbreviation of the chemical name *nicotinamide adenine dinucleotide* (Fig. 4–5). This electron-carrying coenzyme has long been known by the name of *diphosphopyridine nucleotide* (abbreviated DPN). Recently, by international agreement, the former name has been officially adopted because it is chemically more accurate. NAD occurs in all cells and its role is analogous to that of ATP: ATP is a phosphate carrier and NAD is an electron carrier. We will not concern ourselves in detail with its chemical structure, except to note one interesting and important fact. The NAD molecule contains as one of its building blocks the substance *nicotinamide*. It is the nicotinamide portion of the molecule that can accept electrons and become reduced. Nicotinamide is the amide of *nicotinic acid*, a vitamin of the B class, which must be present in the diet of man and other vertebrates. In its absence the dietary deficiency disease *pellagra* results. We can see at once a possible reason for the pathology caused by the lack of nicotinic acid in the diet; if it is not supplied, then the cell is unable to form the complete NAD molecule for lack of an essential building block. Accordingly it is not surprising that in pellagra there is a defect in certain enzymatic reactions involving electron transfers to and from NAD.

4–7 THE PATHWAY OF PHOSPHATE GROUPS IN GLYCOLYSIS

Now we come to the crux of glycolysis, the mechanism by which ATP is generated from ADP. We have already seen in this chapter that the enzymatic oxidation of 3-phosphoglyceraldehyde by NAD proceeds together with the uptake of phosphate, so that the reaction product is the high-energy derivative 1,3-diphosphoglycerate, which in turn donates the 1-phospho group to ADP to form ATP. The elucidation of the mechanism of this reaction by

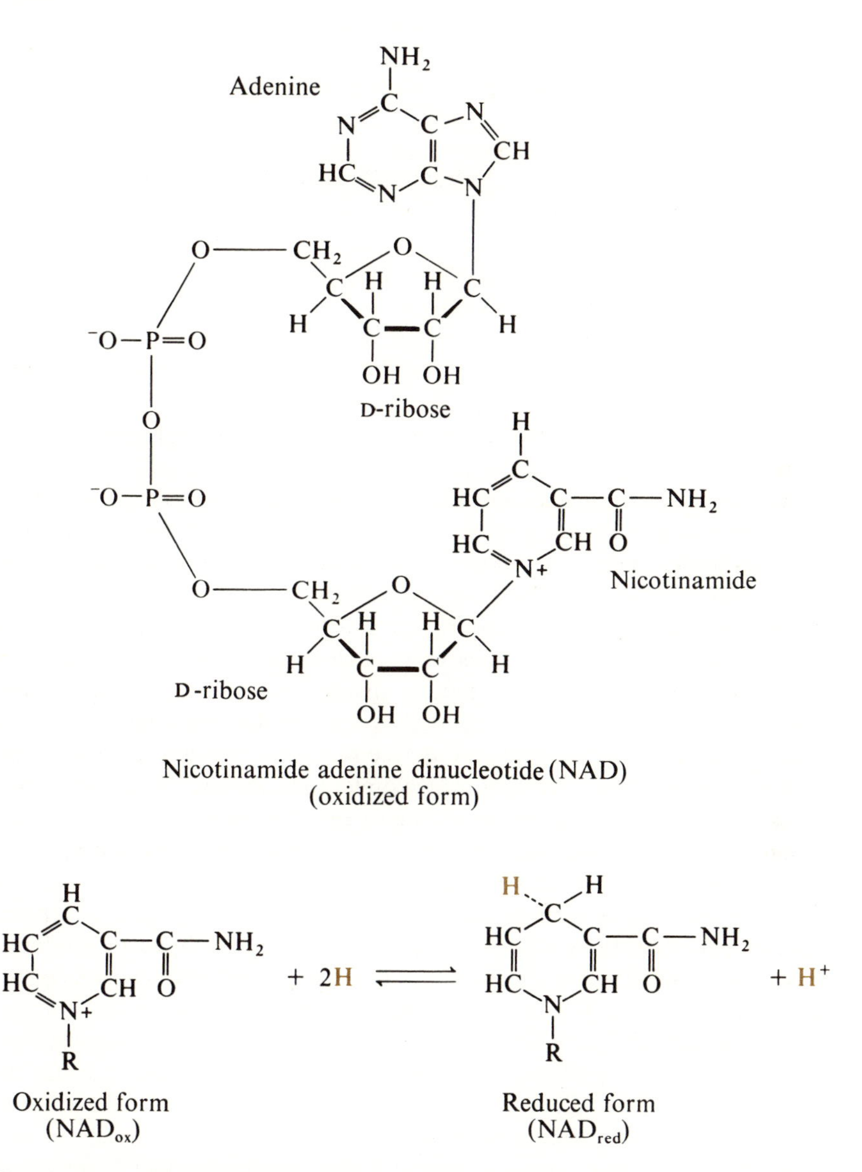

Figure 4–5. Nicotinamide adenine dinucleotide and its action as a hydrogen (electron) carrier.

Warburg in 1939 is a great milestone in modern biology, because it showed for the first time how chemical energy can be recovered from an enzymatic oxidation in the form of ATP. The reaction equations are

$$\text{3-phosphoglyceraldehyde} + P_i + NAD_{ox} \longrightarrow \text{1,3-diphosphoglycerate} + NAD_{red} \quad (4\text{–}4)$$

$$\text{1,3-diphosphoglycerate} + ADP \rightleftharpoons \text{3-phosphoglycerate} + ATP \quad (4\text{–}5)$$

These two reactions illustrate a very crucial and fundamental point common to all known energy-conserving oxidation-reactions in the cell, namely, that oxidation is accompanied by the formation of a high-energy derivative of the oxidation product, which ultimately causes a phosphate group to be donated to ADP. In the reactions given in Eqs. 4–4 and 4–5 the common intermediate (shaded) is the high-energy oxidation product 1,3-diphosphoglycerate. This compound has such a high standard free energy of hydrolysis, namely −11.8 kcal, because its structure contains in rather close juxtaposition two phosphate groups, each having a double negative charge, as well as an unusually high density of electrons in the oxygen-rich anhydride linkage between the carboxyl group and the phosphate group (see Fig. 3–3). This is a rather unstable configuration for these electrons, which prefer to be far away from each other. After hydrolysis of this compound, the forces of repulsion between the two negatively charged products prevent them from recombining again. Moreover, the electrons of the anhydride linkage sink into more stable positions in the carboxyl and phosphate groups. Because of its very high phosphate transfer potential, the carboxyl phosphate group of 1,3-diphosphoglycerate is donated to ADP with a very high thermodynamic driving force, namely, the difference between −11.8 kcal and −7.3 kcal, or about −4.5 kcal.

Because ultimately two molecules of 3-phosphoglyceraldehyde are oxidized by two molecules of pyruvate, two moles of ATP are formed by this oxidation-reduction per mole of glucose entering the cycle.

We have made the point repeatedly that in the glycolytic sequence, as well as all other fermentation sequences, it is the oxidation-reduction reactions that are responsible for generating compounds with high-energy phosphate bonds, which can then donate their phosphate enzymatically to ADP. However, if we inspect the second site of ATP formation in the glycolytic sequence we may be puzzled, because at first glance there appears to be no oxidation-reduction involved at all in the formation of phosphopyruvate from its precursor 2-phosphoglycerate, which is formally an elimination reaction in which water is lost

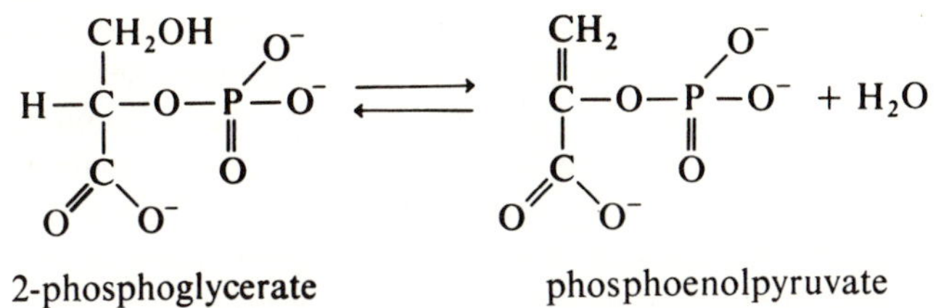

However, there has been a shift of electrons within the molecule in this reaction; it may therefore be regarded as an intramolecular oxidation-reduction reaction. Electrons become more concentrated about carbon atom 2 of the product phosphoenolpyruvate, not only because a double bond is formed (which involves two pairs of electrons) but also by the constriction of the bond distances and angles around this carbon atom. This has the effect of pulling the negatively charged phosphate and carboxyl groups closer together, which, in turn, causes a shift in the internal energy of the molecule, so that phosphopyruvate becomes a high-energy compound ($\Delta G^{0\prime}$ of hydrolysis = −14.8 kcal; see Table 3–1). Phosphoenolpyruvate then donates its phosphate group to ADP in the next reaction, catalyzed by the enzyme *pyruvate kinase*

$$\text{phosphoenolpyruvate} + \text{ADP} \rightleftharpoons \text{pyruvate} + \text{ATP}$$

This reaction goes far toward completion as written because the phosphate transfer potential of phosphopyruvate is much higher than that of ATP (Table 3–1; Fig. 3–4).

We have now seen how chemical energy is conserved in the form of high-energy phosphate groups and thus as ATP in two reactions in the glycolytic scheme. All of the other steps in glycolysis proceed with a relatively small change in free energy and are incapable of delivering ATP.

4–8 THE BALANCE SHEET OF GLYCOLYSIS

Now let us add up all the reactants going into the scheme and all the products emerging to see if the overall balance sheet agrees with Eq. 4–2. One glucose molecule enters at the beginning of the sequence. Then we see that two ATP molecules also enter. One of these brings about the phosphorylation of glucose to glucose 6-phosphate, catalyzed by the enzyme *hexokinase*, and the second is then used to phosphorylate fructose 6-phosphate to form fructose 1,6-diphosphate, a reaction catalyzed by the enzyme *phosphofructokinase*. It may seem anomalous that during glycolysis two molecules of ATP should be cleaved to form ADP, when we have just said that the whole point of glycolysis is to generate ATP from ADP and phosphate. However, the input of two ATP molecules is just a means of priming the pump; these two molecules of priming ATP will be recovered again. Now notice that later in the scheme, two molecules of ADP, two molecules of phosphate, and two molecules of NAD_{ox}, enter at the step in which two molecules of 3-phosphoglyceraldehyde are oxidized to 3-phosphoglycerate (Steps 6 and 7; Fig. 4–3). Still later two more molecules of ADP enter to accept phosphate groups from two molecules of phosphoenolpyruvate (Step 10; Fig. 4–3). The two molecules of NAD_{red} that were formed when 3-phosphoglyceraldehyde was oxidized now return and are utilized in the reduction of pyruvate to lactate.

Now let us see what comes out of the whole process. Two molecules of the 3-carbon lactic acid are formed at the end, corresponding to the six carbon atoms of glucose that were fed into the cycle. Then we see that two molecules of ADP were formed, one at each of the two ATP-requiring priming reactions (Steps 1 and 3; Fig. 4–3). Next, we find that a total of four molecules of ATP were formed, two for each of the two molecules of 3-phosphoglycerate formed (Step 7; Fig. 4–3), and two for each of the two molecules of pyruvate that appeared (Step 10; Fig. 4–3). Now let us write the complete equation for glycolysis, entering into it all the components that went in and all that emerged

$$\text{glucose} + 2\text{ATP} + 2\text{P}_i + 4\text{ADP} + 2\text{NAD}_{ox} + 2\text{NAD}_{red} \longrightarrow 2\text{ lactate} + 2\text{ADP} + 4\text{ATP} + 2\text{NAD}_{red} + 2\text{NAD}_{ox} + 4\text{H}_2\text{O}$$

We may now cancel the components appearing on both sides of the equation and we have the net statement

$$\text{glucose} + 2\text{P}_i + 2\text{ADP} \longrightarrow 2\text{ lactate} + 2\text{ATP} + 2\text{H}_2\text{O}$$

This is identical with Eq. 4–2, the overall equation of anaerobic glycolysis. The two molecules of ATP that were put into the scheme as priming agents were recovered again, and two additional ATP's were formed from ADP and phosphate.

It may appear that the glycolytic sequence is an unnecessarily complex mechanism for the goal achieved. But there are many reasons to think that it represents the simplest possible way in which the glucose molecule can be degraded at pH 7.0 in dilute solution, at one atmosphere pressure and at temperatures compatible with life, in such a manner that a large part of its energy can be conserved in the form of ATP energy. The glycolytic sequence of enzymes was selected over millions of years of cellular evolution and it very likely represents the survival of the fittest molecules. The glycolytic sequence inherently contains highly developed chemical and engineering wisdom, which the biochemist and biophysicist have really just begun to fathom and appreciate.

4–9 INTRACELLULAR ORGANIZATION OF THE GLYCOLYTIC SYSTEM

All of the intermediates in the glycolytic sequence are esters of phosphoric acid; only the initial reactant glucose and the final products pyruvate and lactate are not phosphorylated compounds. The phosphate group, of course, has a very special significance in glycolysis because it is the chemical vehicle required for regenerating ATP from ADP. However, there is another important biological property of phosphorylated compounds. Glucose and its end-product lactate are rather freely permeable through the membranes of cells; glucose enters readily from the outside medium and lactate readily departs from the cell as end product, at least under anaerobic conditions. In contrast, the phosphate ester intermediates of the glycolytic cycle are unable to penetrate through cell membranes; they are in effect locked inside the cell and cannot diffuse out. If such intermediates could escape from the cell, the rate of glycolysis might, of course, become vanishingly small. Moreover, ATP and ADP are also unable to pass through the cell membrane. In general all phosphorylated compounds have a high density of electrical charge at their phosphate groups. The lipid-rich cell membrane resists passage of such highly charged solute molecules. It would appear, then, that phosphate compounds were selected as intermediates during the course of evolution for more than one property.

The eleven enzymes which catalyze the glycolytic sequence exist free in solution in the soluble portion of the cytoplasm, at least in most cells. They are apparently not grouped or arranged in an intracellular structure, as is the case for the much more complex enzyme systems responsible for respiration and photosynthesis, which we shall see are organized in the mitochondria and the chloroplasts respectively.

Finally, it must be pointed out that nearly all the enzyme-catalyzed reactions of glycolysis are freely reversible; that is, they have a relatively small standard free energy change. Later we shall see that the reversible steps of glycolysis are utilized in the reverse direction, in the synthesis of glucose from pyruvic or lactic acid. However, there are three irreversible reactions in the breakdown of glucose to pyruvic acid, those catalyzed by hexokinase, phosphofructokinase, and pyruvate kinase. We shall see later (Section 7–5) that when pyruvic acid is converted into glucose, these irreversible steps are by-passed by other enzymes.

4–10 REGULATION OF THE RATE OF FERMENTATION

Does fermentation proceed at maximal rates at all times or is it regulated? If it is regulated, what factors control its rate?

From much recent research it has become clear that all metabolic pathways are under constant minute-to-minute regulation, so that all components of the metabolic network in the cell are working together harmoniously, and no system is overproducing or underproducing any particular end product. Moreover, it has been found that the regulation of each multienzyme system, such as that catalyzing glycolysis, is brought about by the action of one or more specific *regulatory* enzymes. A regulatory enzyme not only catalyzes a specific reaction in a multienzyme sequence, but it has a second property not present in most enzymes: its catalytic activity is sensitive to the concentration of some crucial metabolite, not necessarily related to its substrate, which can act as a modulator.

If the modulator inhibits the enzyme, it is a negative modulator; if it stimulates the catalytic activity, it is a positive modulator. Regulatory enzymes have been found to possess two types of binding sites, one type for the substrate molecule, called the *catalytic site*, and the other type for the modulator molecule, the *modulator or effector site*. For this reason regulatory enzymes are also called *allosteric* ("other space; other site") enzymes. Regulatory enzymes appear to occur in two molecular forms, one of which is inactive, the other active. The two forms are normally in equilibrium

$$\text{inactive form} \underset{\text{negative modulator}}{\overset{\text{positive modulator}}{\rightleftharpoons}} \text{active form}$$

but the position of the equilibrium is shifted by the presence of the modulator. If a negative modulator is present this equilibrium is shifted to the left, since the negative modulator can combine with the inactive form. Conversely, if a positive modulator is present, the equilibrium is shifted to the right to yield more of the active form, to which the positive modulator is bound. Thus the balance between the active and inactive forms of a regulatory enzyme is determined by the relative concentrations of the positive and negative regulators, which may bear no structural relationship to the substrate. Sometimes, however, the substrate or product of a regulatory enzyme may also act as a modulator, as we shall see.

The sequence of eleven enzymes responsible for anaerobic glycolysis contains such a regulatory enzyme, namely, phosphofructokinase, which catalyzes the phosphorylation of fructose 6-phosphate to fructose 1,6-diphosphate

$$\text{ATP} + \text{fructose 6-phosphate} \longrightarrow \text{ADP} + \text{fructose 1,6-diphosphate}$$

The rate of this enzymatic reaction has been found to be profoundly influenced by ATP and ADP in a manner apart from the usual dependence of an enzyme reaction rate on its substrates or products. Although this enzyme requires ATP as a substrate, its rate in the forward direction is greatly accelerated if ADP is also present in high concentrations. Conversely, if the ATP concentration becomes significantly higher than is required to saturate the catalytic site of the enzyme, then the catalytic rate is diminished. ADP is thus a positive modulator and ATP a negative or inhibitory modulator of phosphofructokinase, as can be represented in the equation

$$\underset{\text{(inactive form)}}{\text{phosphofructokinase}} \underset{\text{ATP}}{\overset{\text{ADP}}{\rightleftharpoons}} \underset{\text{(active form)}}{\text{phosphofructokinase}}$$

The rate of the reaction catalyzed by phosphofructokinase is therefore regulated by the two major components of the energy transfer system of the cell, namely ADP and ATP. The sum of the concentrations of ADP and ATP in the cell is always constant. Therefore, whenever this system is filled with high-energy phosphate groups, that is, when all the available ADP in the cell has been phosphorylated to ATP, the high ATP/ADP ratio causes this equilibrium to swing to the left, favoring the formation of the catalytically inactive form of the enzyme. If, however, some of the ATP now suddenly undergoes dephosphorylation in an energy-requiring process, the ADP concentration in the cell would increase and that of ATP would decrease. The rate of the phosphofructokinase reaction would then rise, since more of the active form of the enzyme would be formed, as a consequence, the rate of glycolysis would increase and more ATP would be generated to replace that dephosphorylated in the energy-requiring process.

We see that ATP, an end product of glycolysis or fermentation, is a feedback inhibitor of the system that produces it. On the other hand, lactic acid (or ethanol + CO_2), the other products of anaerobic fermentation, are not modulators of the action of phosphofructokinase.

4–11 THE ENERGETICS OF FERMENTATION AND RESPIRATION COMPARED

Now we come to another interesting point that arises from considering the amounts of energy that anaerobic or fermentative degradation of glucose can yield, in comparison with the energy yield when glucose is oxidized to CO_2 and H_2O. Figure 4–6 shows the standard free energy changes associated with anaerobic conversion of glucose to lactate and with the complete oxidation of glucose to CO_2 and H_2O as it occurs in aerobic cells. Anaerobic cells obtain energy from the conversion of glucose to lactate, which then leaves the cell as waste. However, the maximum energy available in this step is only 47 kcal/ mole of glucose, or a little under 7 per cent of the amount available when glucose is oxidized to CO_2 and H_2O, namely 686 kcal/mole. For the advantage of being able to extract energy from glucose in the absence of oxygen, the anaerobic organism must waste over 93 per cent of the total energy it might be able to obtain if it could oxidize glucose with molecular oxygen to CO_2 and H_2O.

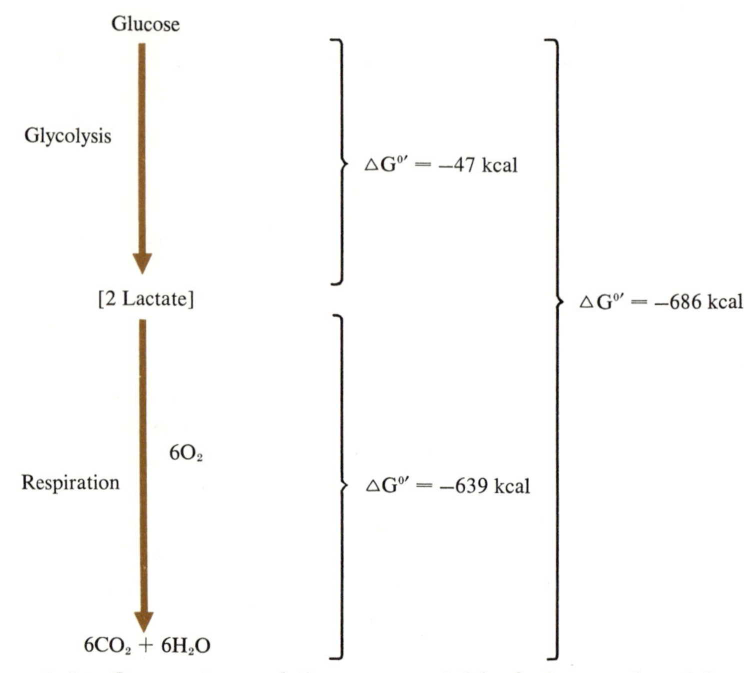

Figure 4–6. Comparison of the energy yield of glucose breakdown in glycolysis vs respiration.

In the aerobic organism lactate does not leave the cell; instead it, or pyruvate, is oxidized to CO_2 and H_2O, with recovery of much of the other 93 per cent of the energy of glucose. Respiration, the oxidation of glucose with molecular oxygen, is clearly very efficient in extracting all the possible energy from the glucose molecule. We will see in the next chapter that respiration is also a much more complex process than glycolysis. Respiration is to glycolysis what a jet engine is to a " one-lung " reciprocating engine. But each type of engine has its specific biological role to fulfill.

Because anaerobic glycolysis can extract only a small fraction of the total energy of the glucose molecule, it is a corollary that anaerobic cells must use much more fuel per unit of time per unit of weight to accomplish the same amount of cellular work as an aerobic cell. It has been found that anaerobic cells may use over ten times as much glucose as aerobic cells to do the same amount of work, and they can consume many times their weight of glucose in only short periods of time. This fact has some interesting consequences for facultative cells. For example, cancer cells, which are facultative cells and can thus live either anaerobically or aerobically, have a metabolic defect which causes them to use up very large quantities of glucose by glycolysis, even though they are supplied with oxygen and are still able to respire.

5

RESPIRATION AND ATP FORMATION IN THE MITOCHONDRION

The real mainspring of energy in aerobic cells is respiration, the enzymatic oxidation of fuel molecules by molecular oxygen. The enzyme systems that catalyze respiration and the conservation of respiratory energy as ATP are far more complex than those concerned in fermentation or glycolysis. They involve many more enzymes and many more separate chemical steps. Furthermore, these enzymes do not occur singly in the free form in the soluble portion of the cell cytoplasm. Rather they are fixed in geometrically specific arrays or assemblies in the *mitochondria*, the "power plants" of the cell. The molecular basis of the structure and function of these energy-converting enzyme assemblies offers a most challenging problem.

In this chapter we shall consider first how the major nutrient molecules are prepared for entry into the oxidative pathways of aerobic cells. Then we shall analyze the three major phases of biological oxidations: (1) the tricarboxylic acid cycle, (2) electron transport, and (3) the recovery of oxidative energy as ATP, in the process called oxidative phosphorylation. The master plan of biological oxidations is shown schematically in Fig. 5–1.

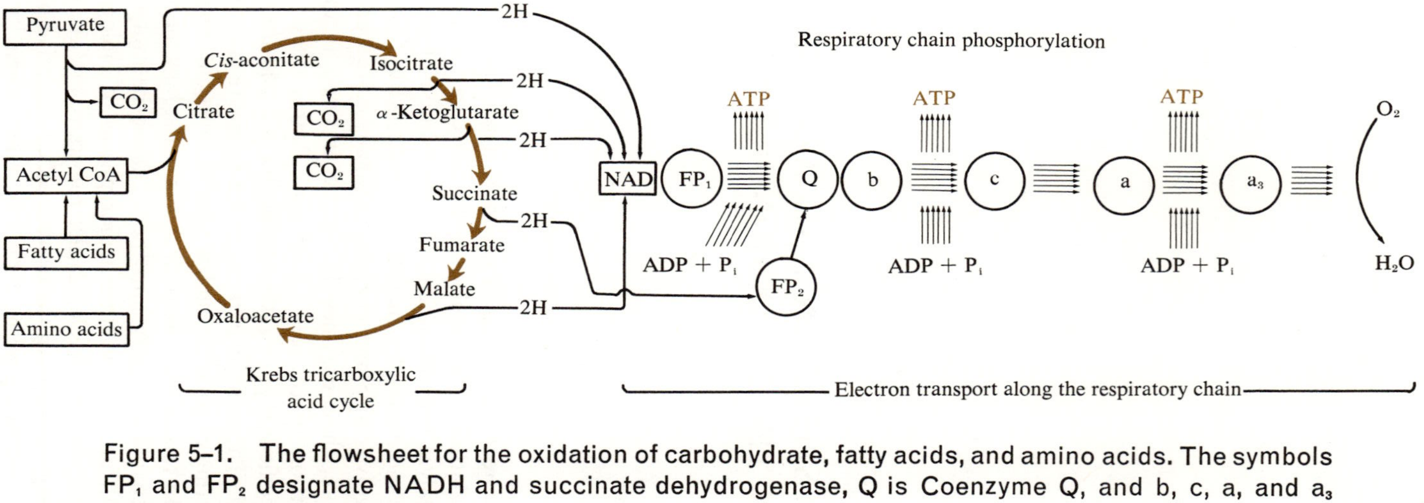

Figure 5–1. The flowsheet for the oxidation of carbohydrate, fatty acids, and amino acids. The symbols FP_1 and FP_2 designate NADH and succinate dehydrogenase, Q is Coenzyme Q, and b, c, a, and a_3 designate the cytochromes.

5-1 THE FORMATION OF ACETYL CoA

All three major foodstuffs of heterotrophic cells, carbohydrate, fatty acids, and amino acids, are ultimately oxidized by the cyclic series of reactions known as the Krebs tricarboxylic acid cycle. However, each of these nutrients must first be prepared for the cycle by preliminary enzymatic reactions in which the carbon skeleton of the foodstuff molecule is broken down into 2-carbon pieces. The tricarboxylic acid cycle can accept as fuel only the 2-carbon compound acetic acid in an activated form, whose structure we shall presently examine.

The pyruvic acid formed as the aerobic end product of the glycolytic reactions is prepared for the cycle by undergoing enzymatic oxidation by the *pyruvic dehydrogenase complex* to a derivative of acetic acid; in this reaction the third carbon atom of pyruvic acid, that of the carboxyl group, is lost as CO_2

$$\underset{\text{pyruvic acid}}{CH_3COCOOH} + NAD_{ox} + CoA\text{—}SH \longrightarrow \underset{\text{acetyl CoA}}{CH_3CO\text{—}S\text{—}CoA} + CO_2 + NAD_{red} \quad (5\text{–}1)$$

The electrons removed from pyruvic acid are accepted by NAD_{ox} and carried in the form of NAD_{red} to the respiratory chain, a process we shall discuss later. The acetic acid formed in this reaction is not in the free form, but appears as a derivative of *coenzyme A*, a complex nucleotide whose structure is shown in Fig. 5–2. Coenzyme A is a carrier of acetyl groups in the cell, just as ATP is a carrier of phosphate groups and NAD is a carrier of electrons. CoA has a structure reminiscent of that of NAD and ATP because it contains adenine, ribose, and a pyrophosphate bridge. In addition, it contains the compound *pantothenic acid*, a vitamin of the B complex. The thiol (—SH) group of CoA serves to bind acetic acid covalently through a thioester linkage, which is actually a high-energy bond. The standard free energy of hydrolysis of acetyl CoA

$$CH_3CO\text{—}S\text{—}CoA + H_2O \longrightarrow CH_3COOH + CoA\text{—}SH$$

at pH 7.0 is $\Delta G^{0\prime} = -7.5$ kcal, which is somewhat higher than the standard free energy of hydrolysis of ATP. The acetyl group carried by CoA—SH is thus in an activated or energized form, and it may be transferred enzymatically to various acetyl group acceptors in the course of cell metabolism. The acetyl group of acetyl CoA is the immediate fuel of the tricarboxylic acid cycle, and is made available to the cycle by an enzymatic transfer reaction.

Fatty acids and amino acids also yield acetyl CoA as major degradation

Acetyl CoA

Pantothenic acid (a vitamin)

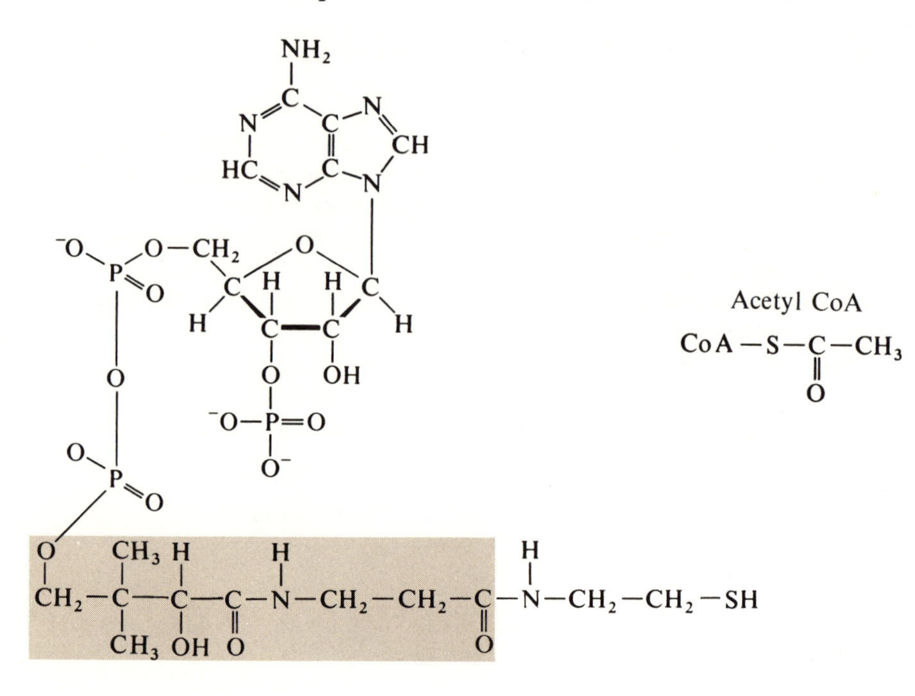

Figure 5–2. Coenzyme A and acetyl coenzyme A.

products. Figure 5–3 shows how palmitic acid, a 16-carbon fatty acid that is representative of the major components of lipids, is degraded enzymatically to yield eight molecules of acetyl CoA. In this process we see still another function of coenzyme A: all the intermediate steps of fatty acid oxidation occur in the form of thioesters of the intermediates with coenzyme A. The pathways by which the twenty different amino acids are degraded to form acetyl CoA are varied and complex; we shall not examine them in detail. Ultimately, however, most of the carbon atoms of carbohydrates, fatty acids, and amino acids finally appear as acetyl CoA and may then enter the tricarboxylic acid cycle (Fig. 5–1).

Figure 5–3. Oxidation of a fatty acid. Successive acetyl residues are removed from the carboxyl end of the palmitic acid molecule by a sequence of four reactions (not shown), to yield ultimately eight molecules of acetyl CoA. The carbon atoms are numbered to show their origin from palmitic acid.

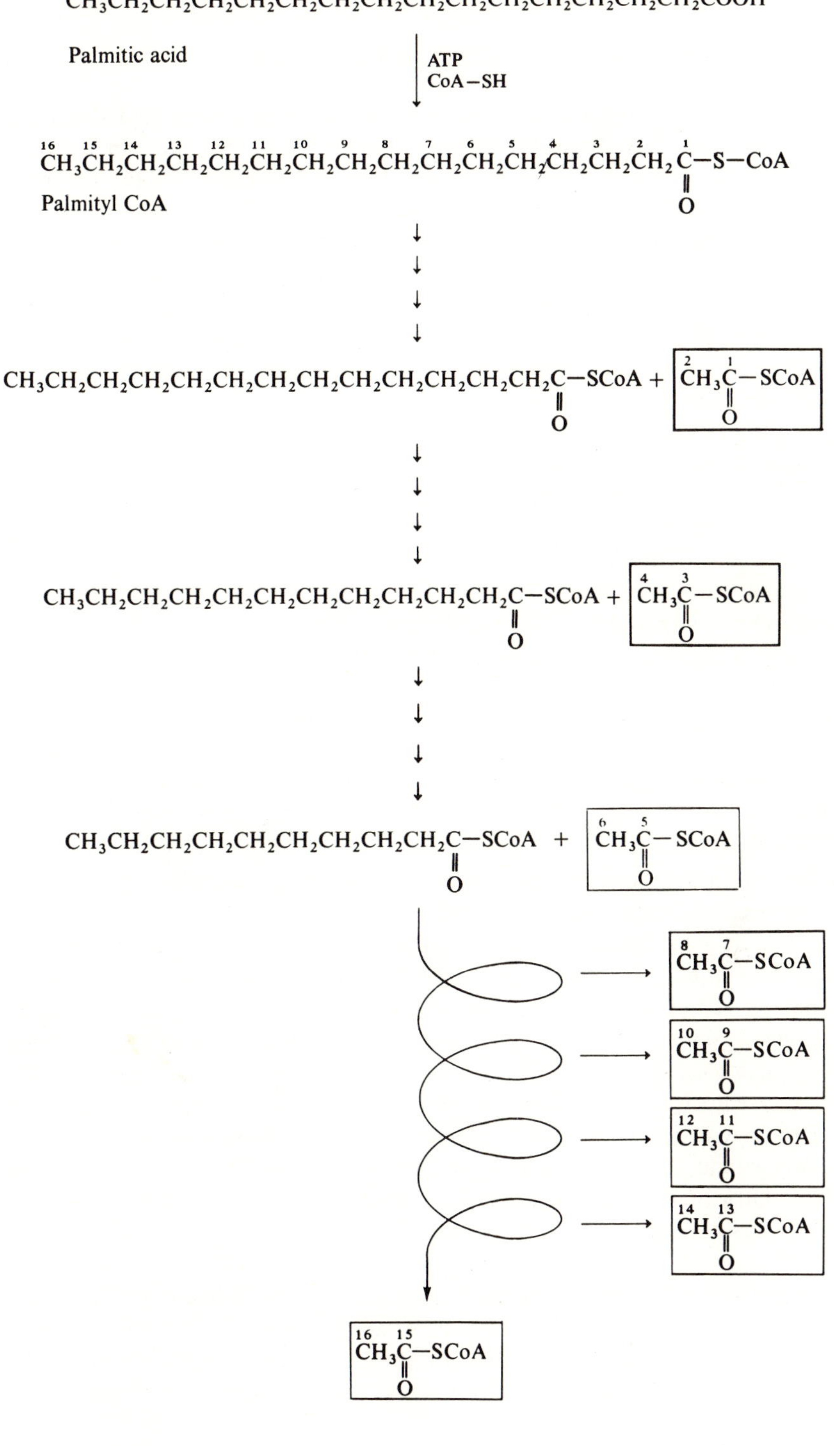

CH3CH2CH2CH2CH2CH2CH2CH2CH2CH2CH2CH2CH2CH2CH2COOH
Palmitic acid
ATP
CoA–SH
CH3CH2CH2CH2CH2CH2CH2CH2CH2CH2CH2CH2CH2CH2CH2C–S–CoA
Palmityl CoA
CH3CH2CH2CH2CH2CH2CH2CH2CH2CH2CH2CH2CH2C–SCoA + CH3C–SCoA
CH3CH2CH2CH2CH2CH2CH2CH2CH2CH2CH2C–SCoA + CH3C–SCoA
CH3CH2CH2CH2CH2CH2CH2CH2CH2C–SCoA + CH3C–SCoA
CH3C–SCoA
CH3C–SCoA
CH3C–SCoA
CH3C–SCoA
CH3C–SCoA

5–2 THE TRICARBOXYLIC ACID CYCLE

This cyclic series of enzymatic reactions has an interesting experimental history. In the middle 1930's Szent-Györgyi in Hungary found that succinic, fumaric, malic, and oxaloacetic acids, which have four carbon atoms and two carboxyl groups, are vigorously oxidized by suspensions of minced muscle. However, he found they have a second action: they greatly stimulate the oxidation of carbohydrate by muscle. Moreover, Szent-Györgyi found that this effect is catalytic, that is, for each molecule of succinic acid added to the muscle, many molecules of muscle carbohydrate were oxidized. Parenthetically, we may say at this point that these compounds were tested, not as free acids, but after neutralization to the pH of the muscle suspensions, approximately pH 7.0. At this pH these organic acids exist as their anions, namely, as succinate, fumarate, malate, and oxaloacetate. We shall often have occasion to refer to various organic acids as intermediates in metabolism; it will be understood of course that they will usually exist as their anions at the pH of intracellular fluid.

A few years later, Krebs, a British biochemist, surveyed a variety of other naturally occurring dicarboxylic and tricarboxylic organic acids to see whether they also would show the catalytic effect exhibited by the four acids studied by Szent-Györgyi. He made the important observation that there were three 6-carbon tricarboxylic acids (citric acid, *cis*-aconitic acid, and isocitric acid) and a 5-carbon dicarboxylic acid (α-ketoglutaric acid) that are also capable of stimulating the aerobic oxidation of carbohydrate in muscle suspensions. But no other organic acids of similar structure produced this effect.

Krebs therefore concluded that the organic acids having this catalytic effect are interrelated and participate together in a metabolic sequence. He arranged them into a series (Fig. 5–4) and proved that each of the postulated enzymatic steps takes place in muscle at a rate consistent with the rate of respiration. But he showed that these enzymatic reactions function in a cycle, and thus differ from the reactions of glycolysis, which take place in a linear sequence. Two points of evidence supported the occurrence of a cyclic series of reactions. First, Krebs found that oxaloacetic acid, the last intermediate in the sequence (Fig. 5–2) reacted enzymatically with a breakdown product of pyruvic acid (now known to be acetyl CoA) to make citric acid, the first intermediate, a step that joined the "head" and the "tail" of the sequence from citric acid to oxaloacetic acid. The citric acid so formed then was oxidized again to yield oxaloacetic acid, which could react with another molecule

---

Figure 5–4. The major reaction steps in the tricarboxylic acid cycle. Malonic acid inhibits succinic dehydrogenase and thus prevents regeneration of oxaloacetic acid. The path of the carbon atoms of the acetyl group is shown in color.

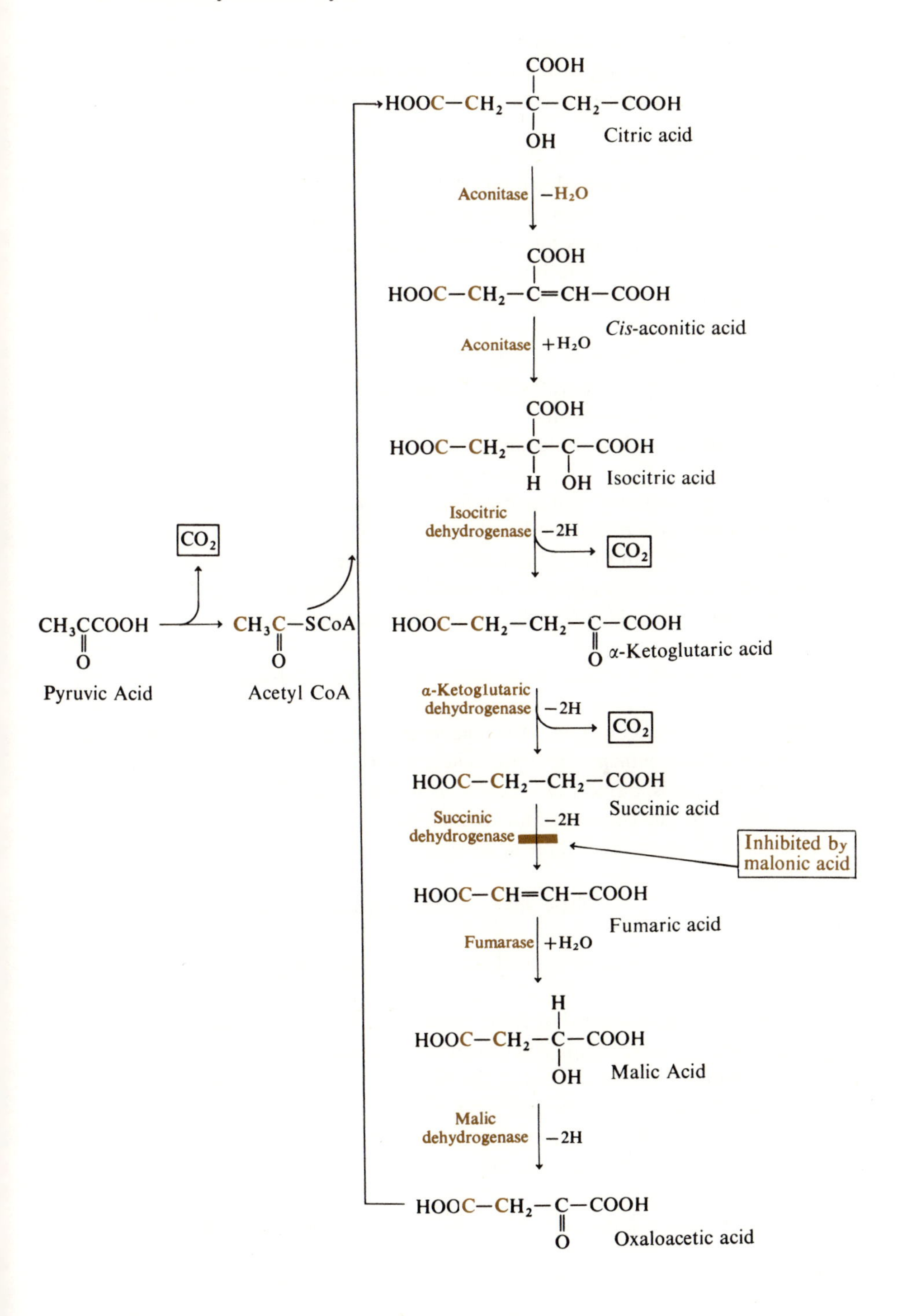
Citric acid
Aconitase
$-H_2O$
Cis-aconitic acid
Aconitase
$+H_2O$
Isocitric acid
Isocitric dehydrogenase
$-2H$
CO_2
CO_2
Pyruvic Acid
Acetyl CoA
α-Ketoglutaric acid
α-Ketoglutaric dehydrogenase
$-2H$
CO_2
Succinic acid
Succinic dehydrogenase
$-2H$
Inhibited by malonic acid
Fumaric acid
Fumarase
$+H_2O$
Malic Acid
Malic dehydrogenase
$-2H$
Oxaloacetic acid

of pyruvic acid to start another cycle. In each cycle three molecules of CO_2 appeared corresponding to the input of one molecule of pyruvic acid. Thus one molecule of oxaloacetic acid could cause oxidation of many molecules of pyruvic acid.

Krebs then showed that the compound *malonic acid*, which is closely related to succinic acid and inhibits the enzyme *succinic dehydrogenase*, completely blocks the utilization of pyruvic acid by muscle. Krebs therefore concluded that these reactions must be linked into a cycle and that if any single reaction of the sequence is blocked, the utilization of pyruvate must stop. This cycle, which is called the *tricarboxylic acid cycle*, *citric acid cycle*, or simply the *Krebs cycle*, has since been found to occur in almost all heterotrophic cells and is the major route of oxidative metabolism of all foodstuffs. That the cycle actually operates in the intact cell has been completely verified by isotopic tracer experiments.

Now, starting with a molecule of acetyl CoA, let us follow through the individual steps of the Krebs cycle with the aid of the structural formulas shown in the schematic plan in Fig. 5–4. Keep in mind that it is the essential purpose of the cycle to dismember the acetyl group of acetyl CoA to yield two molecules of CO_2. In the first reaction of the cycle, the acetyl group of acetyl CoA is enzymatically transferred to the 4-carbon dicarboxylic acid *oxaloacetic acid* to form *citric acid*, a 6-carbon tricarboxylic acid

$$\text{acetyl CoA} + \text{oxaloacetic acid} \longrightarrow \text{citric acid} + \text{CoA—SH}$$

In this reaction free coenzyme A is regenerated. In the next reaction, catalyzed by the enzyme *aconitase*, the citric acid is reversibly converted into two other tricarboxylic acids, namely *cis-aconitic acid* and *isocitric acid*. This conversion takes place by successive removal and addition of water molecules. An equilibrium mixture of all three acids is thus formed

$$\text{citric acid} \rightleftharpoons \textit{cis}\text{-aconitic acid} + H_2O$$

$$H_2O + \textit{cis}\text{-aconitic acid} \rightleftharpoons \text{isocitric acid}$$

In the next step, catalyzed by the enzyme *isocitric dehydrogenase*, two hydrogen atoms are removed from isocitric acid and accepted by the electron carrier NAD_{ox}, which becomes reduced to NAD_{red}. From the isocitric acid are formed the 5-carbon dicarboxylic acid, *α-ketoglutaric acid*, and a molecule of CO_2. This molecule of CO_2 is the first of the two which arise from the 2-carbon acetic acid fed into the cycle

$$\text{isocitric acid} + NAD_{ox} \longrightarrow \alpha\text{-ketoglutaric acid} + CO_2 + NAD_{red}$$

The second CO_2 molecule arises in the following step, catalyzed by *α-ketoglutaric dehydrogenase.* The 5-carbon compound α-ketoglutaric acid is then oxidized to CO_2 and the 4-carbon compound succinic acid, in a complex sequence of reactions in which one molecule of ATP is formed from ADP and phosphate. The CO_2 is derived from a carboxyl group of α-ketoglutaric acid and NAD_{ox} serves as electron acceptor

$$\alpha\text{-ketoglutaric acid} + NAD_{ox} + P_i + ADP \longrightarrow \text{succinic acid} + CO_2 + ATP + NAD_{red}$$

In the next step, *succinic acid* is dehydrogenated to *fumaric acid* by succinic dehydrogenase; the active group of this enzyme accepts the hydrogen atoms and becomes reduced

$$\text{succinic acid} \xrightarrow{-2H} \text{fumaric acid}$$

The *fumaric acid* is now hydrated at the double bond to form *malic acid,* by the action of the enzyme *fumarase*

$$\text{fumaric acid} + H_2O \rightleftharpoons \text{malic acid}$$

Finally, the 4-carbon compound malic acid is dehydrogenated to oxaloacetic acid by *malic dehydrogenase*; again, NAD_{ox} is the electron acceptor

$$\text{malic acid} + NAD_{ox} \rightleftharpoons \text{oxaloacetic acid} + NAD_{red}$$

With the regeneration of a molecule of the 4-carbon compound oxaloacetic acid, which we will recall was the component of the cycle with which the acetyl CoA originally combined, we have completed one revolution of the Krebs cycle. Now let us see what has been accomplished.

We have fed into the cycle one molecule of the 2-carbon compound acetic acid as acetyl CoA, as well as one molecule of the 4-carbon compound oxaloacetic acid. Two atoms of carbon appeared as two molecules of CO_2 and a molecule of the 4-carbon compound oxaloacetic acid was regenerated. These reactions thus describe the fate of the carbon skeleton of acetic acid. Accompanying these reactions of the tricarboxylic acid cycle there were four steps in which pairs of hydrogen atoms were removed by dehydrogenases; we shall return to these presently. But it is of the greatest importance to note again that oxaloacetic acid, which was required to get the fuel into the cycle at its beginning, is regenerated. Oxaloacetic acid can now react with a second molecule of acetic acid and start a new cycle, and again oxaloacetic acid will be regenerated, and so on. One molecule of oxaloacetic acid can therefore bring about the oxidation of an infinite number of acetic acid molecules to CO_2 and H_2O, simply because it is regenerated by the cycle at each turn.

5–3 ELECTRON TRANSPORT AND THE RESPIRATORY CHAIN

We have seen that in each revolution of the tricarboxylic acid cycle there are four dehydrogenation steps. In three of these, NAD_{ox} served as the electron acceptor for the specific dehydrogenases oxidizing isocitric, α-ketoglutaric, and malic acids, respectively; thus three molecules of NAD_{red} were formed in each revolution. In the other oxidation step, the pair of electrons removed from succinic acid was accepted by the active group of succinic dehydrogenase, which is a *flavoprotein*. Flavoproteins are a class of dehydrogenases containing as electron acceptor a flavin nucleotide. Succinic dehydrogenase contains *flavin adenine dinucleotide*, abbreviated FAD, which is an electron carrier similar to NAD in its action. Flavin nucleotides are also noteworthy in having as a building block the vitamin *riboflavin* or *vitamin* B_2 (Fig. 5–5). The three NAD_{red} molecules and the FAD_{red} formed in the cycle now donate their electrons to another series of enzymes that constitute the *respiratory chain*, as we can see in Fig. 5–1.

The respiratory chain is the *final common pathway* by which *all* electrons derived from different fuels of the cell flow to oxygen, the final oxidant or acceptor of electrons in aerobic cells. Electron transport along this chain is the real " business end " of all cellular oxidations because the electrons entering the respiratory chain have a relatively high energy content. However, as they flow down the chain they lose much of their energy, which we shall presently see is conserved in the form of ATP.

The respiratory chain (Fig. 5–1) is not yet completely understood in detail. However, it appears certain that the electrons from NAD_{red} and FAD_{red} funnel into the common acceptor *coenzyme Q* (Fig. 5–5) and then into a series of *cytochromes*, which are electron-transferring enzyme molecules containing active groups called *hemes*, which consist of porphyrin and iron. Cytochromes are red in color and resemble structurally the oxygen-carrying pigment *hemoglobin* of the red blood cell. The iron atom of each cytochrome molecule can exist in the Fe(II) or ferrous form, as well as in the Fe(III) or ferric form. Thus each cytochrome in its Fe(III) form can accept an electron and become reduced to the Fe(II) form. This in turn can donate its electron to the next carrier, in its oxidized form, and so on. The cytochromes differ somewhat chemically and in molecular weight; each can be recognized by its characteristic absorption spectrum. Only the last one, *cytochrome* a_3 or *cytochrome oxidase*, can give up its electrons to molecular oxygen directly.

The transfer of electrons down the respiratory chain can be written as a

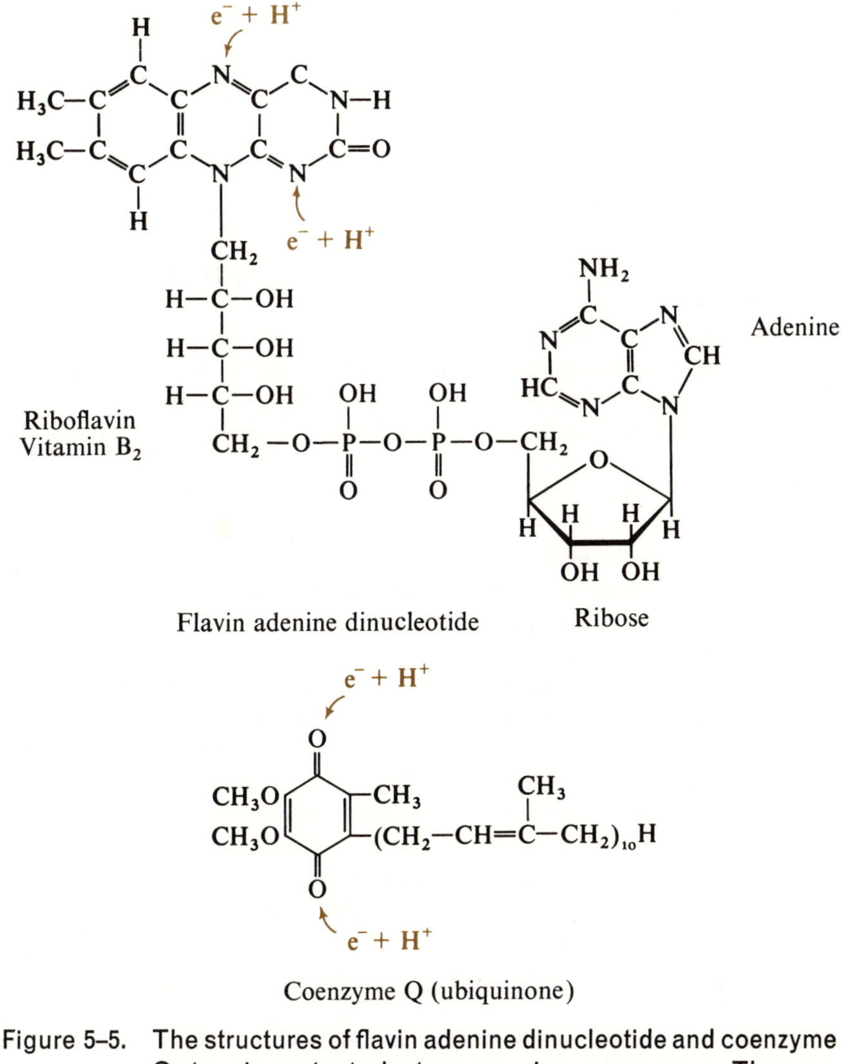

Figure 5–5. The structures of flavin adenine dinucleotide and coenzyme Q, two important electron carrying coenzymes. They are shown in their oxidized forms. The points at which they accept hydrogen (electrons + H^+) are shown by arrows in color.

series of consecutive reactions connected by common intermediates. For the chain from NAD_{red} to oxygen, we may write

$$NAD_{red} + flavoprotein_{ox} \longrightarrow NAD_{ox} + flavoprotein_{red} \quad (5\text{–}2)$$

$$flavoprotein_{red} + CoQ_{ox} \longrightarrow flavoprotein_{ox} + CoQ_{red} \quad (5\text{–}3)$$

$$CoQ_{red} + 2\ cyt\ b_{ox} \longrightarrow CoQ_{ox} + 2\ cyt\ b_{red} + 2H^+ \quad (5\text{–}4)$$

$$2\ cyt\ b_{red} + 2\ cyt\ c_{ox} \longrightarrow 2\ cyt\ b_{ox} + 2\ cyt\ c_{red} \quad (5\text{–}5)$$

$$2\ cyt\ c_{red} + 2\ cyt\ a_{ox} \longrightarrow 2\ cyt\ c_{ox} + 2\ cyt\ a_{red} \quad (5\text{–}6)$$

$$2\ cyt\ a_{red} + 2\ cyt\ a_{3\,(ox)} \longrightarrow 2\ cyt\ a_{ox} + 2\ cyt\ a_{3\,(red)} \quad (5\text{–}6a)$$

$$2\ cyt\ a_{3\,(red)} + \tfrac{1}{2}O_2 + 2H^+ \longrightarrow 2\ cyt\ a_{3\,(ox)} + H_2O \quad (5\text{–}7)$$

The cytochromes can carry only one electron at a time, whereas the NAD and flavoprotein can carry two at a time. For this reason, each cytochrome is shown as reacting twice. The overall reaction for oxidation of NAD_{red} by molecular oxygen then is the sum of reactions (5–2) through (5–7), namely

$$NAD_{red} + \tfrac{1}{2}O_2 + 2H^+ \longrightarrow NAD_{ox} + H_2O \quad (5\text{–}8)$$

Thus for each pair of hydrogen atoms that are removed at each of the four dehydrogenation steps of the Krebs cycle and become a pair of H^+ ions, a pair of electrons enter the respiratory chain and ultimately reduce one atom of oxygen to form water.

Finally, we must add one other important fact. It has been found that the enzymatic transport of electrons to oxygen can be inhibited by certain poisons. For example, cyanide in very low concentrations inhibits completely the transfer of electrons to oxygen by cytochrome *a*. This is why cyanide is one of the most toxic poisons known; it blocks virtually all the biological oxidations in the body because it poisons the last step in the respiratory chain. Other characteristic inhibitors of electron transport are *rotenone*, a naturally occurring toxic substance used as an insecticide, and *antimycin A*, an antibiotic.

5–4 THE ENERGETICS OF ELECTRON TRANSPORT

When electrons are transferred from one compound to another, an oxidation-reduction reaction takes place. Each electron donor or reducing agent shows a characteristic *electron pressure* and each electron acceptor or oxidizing agent a characteristic *electron affinity*. Such electron pressures or affinities can be measured in terms of an electromotive force or potential. Each electron donor, tested under standard conditions, has a characteristic *standard*

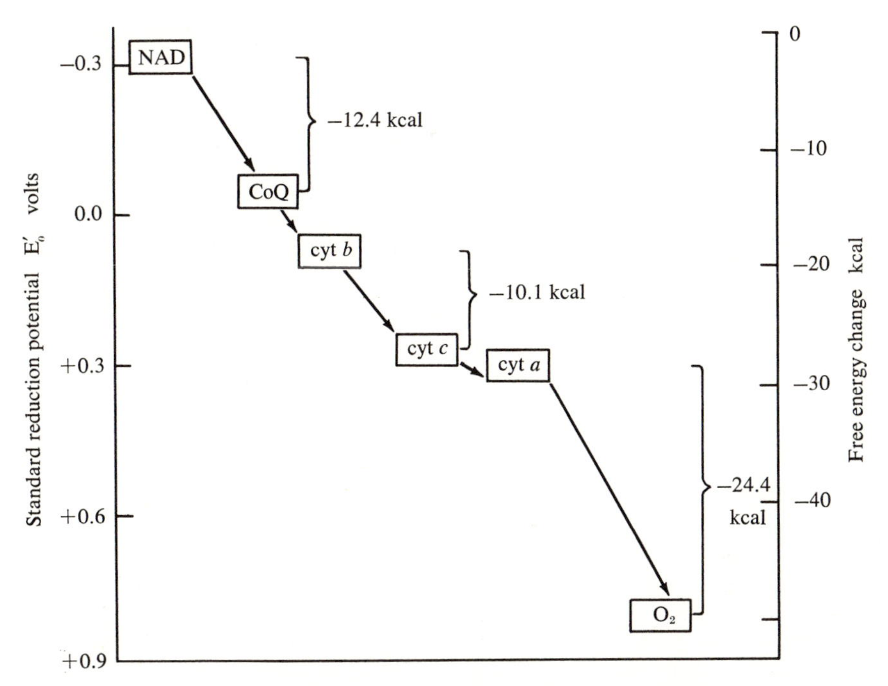

Figure 5–6. Release of energy as electrons flow from NAD_{red} to oxygen. At three points the standard free energy decrease is especially large.

reduction potential, symbolized E_0'. Electron donors may therefore be arranged in a thermodynamic series of decreasing electron pressures, just as phosphate compounds can be arranged in a series of decreasing phosphate transfer potentials, as we saw in Section 3–5. In such a series, it will be the tendency for electrons to flow from the most negative compound, that having the highest electron pressure, to more positive compounds lower in the scale. In Fig. 5–6 is shown such a vertical scale of standard reduction potentials of each of the electron carriers of the respiratory chain. Electrons will tend to pass from the most negative carrier (NAD) to the more positive carriers below it. We see that the electron carriers of the respiratory chain thus permit a stepwise flow of electrons to increasingly more positive acceptors until they finally meet oxygen, the most electropositive acceptor, which is thus reduced to water.

The decline in free energy at each electron transfer is directly related to the magnitude of the decrease in electron pressure by the equation

$$\Delta G^{0\prime} = -nF\Delta E_0'$$

where $\Delta G^{0\prime}$ is the standard free energy change, n is the number of electrons, F is the Faraday (23,040 cal/volt), and $\Delta E_0'$ is the difference in the standard reduction potentials of the reacting carriers. This equation is formally similar to that developed in Section 2–7, namely

$$\Delta G^{0\prime} = -RT \ln K_{eq}$$

Figure 5–6 shows the decrease in free energy that occurs as a pair of electrons moves from each carrier of the respiratory chain to the next. We note that there are three relatively large drops and two relatively small ones; we will come back to this point later. For the moment we will concern ourselves with the magnitude of the free energy change as a pair of electrons moves the entire length of the chain, from NAD to oxygen, in the overall reaction

$$NAD_{red} + \tfrac{1}{2}O_2 + 2H^+ \longrightarrow NAD_{ox} + H_2O$$

This reaction is seen to proceed with an extremely large decrease in free energy ($\Delta G^{0\prime} = -52$ kcal); in fact, sufficient energy is released to make several molecules of ATP from ADP and phosphate.

If we turn back to Fig. 5–1 and sum up all the dehydrogenation steps, we see that altogether twelve pairs of electrons pass down the respiratory chain to oxygen during complete oxidation of one molecule of glucose to CO_2 and H_2O. The process of electron transport from NAD_{red} to oxygen can thus account for about $12 \times 52{,}000 = -624$ kcal/mole of glucose oxidized. If we recall that the free energy of combustion of glucose is -686 kcal/mole, it is clear that almost all the free energy decrease in the biological oxidation of glucose occurs during the enzymatic transport of electrons from the first electron acceptor down the respiratory chain to molecular oxygen.

5–5 OXIDATIVE PHOSPHORYLATION

Careful quantitative studies of electron transport in intact, isolated mitochondria, in which all the enzymatic reactions of the Krebs cycle and electron transport take place, has revealed that both phosphate and ADP are necessary components for maximum rates of electron transport. Furthermore, they are used up in the process and ATP is formed. In fact the important finding was made in 1951 that when a single pair of electrons travels from NAD_{red} to oxygen along the chain, not just one, but three molecules of ATP are formed from ADP and phosphate. Such phosphorylation in the respiratory chain is called *oxidative phosphorylation* or *respiratory chain phosphorylation*. Therefore we must conclude that the true equation of electron transport in intact mitochondria is not

$$NAD_{red} + \tfrac{1}{2}O_2 + 2H^+ \longrightarrow NAD_{ox} + H_2O \qquad (5\text{–}8)$$

but rather the following

$$NAD_{red} + 3ADP + 3P_i + 2H^+ + \tfrac{1}{2}O_2 \longrightarrow NAD_{ox} + 3ATP + 4H_2O \qquad (5\text{–}9)$$

Because formation of three moles of ATP requires input of at least $3 \times 7.3 = 21.9$ kcal and the oxidation of NAD_{red} (Eq. 5–8) delivers 52 kcal, we can deduce that the oxidative phosphorylation of three moles of ADP conserves $(21.9/52)100 = 42$ per cent of the total energy yield when one mole of NAD_{red} is oxidized by oxygen. Respiratory chain phosphorylation is therefore the major mechanism for the conservation of the large amounts of energy made available during the aerobic phase of glucose oxidation.

If we now return to Fig. 5–6, we will see that there are three segments of the chain in which there is a relatively large free energy decrease: from NAD to coenzyme Q, from cytochrome *b* to cytochrome *c*, and from cytochrome *a* to oxygen. It is at these points in the respiratory chain that high-energy intermediates or states are generated during electron transport, which ultimately can donate high-energy phosphate groups to ADP to yield ATP. We can now understand why it is a biological necessity for the respiratory chain to have many members acting in sequence, and not just one or two. Oxidation of NAD_{red} by oxygen proceeds with a very large free energy drop of some 52 kcal, whereas the standard biological energy currency is in the form of packets of 7.3 kcal, equivalent to the standard free energy of formation of ATP from ADP and phosphate. The respiratory chain is thus a molecular device for delivering energy in a series of small packets, three of which are energetically equivalent to ATP. This is made possible by lowering the energy of the electrons gradually in a series of small steps, as we have seen in Fig. 5–6.

Respiratory chain phosphorylation can be inhibited by certain poisons in such a way that electron transport still continues but the linked phosphorylation of ADP to ATP does not. Under these conditions the energy of oxidation of glucose is dissipated completely as heat and none is recovered as ATP. Such poisons are called *uncoupling agents*. An example is 2,4-dinitrophenol:

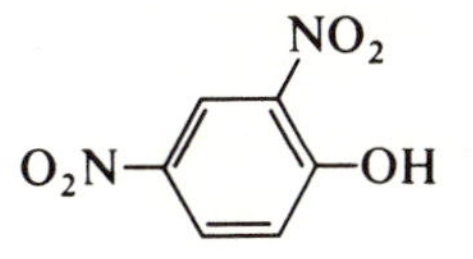

When 2,4-dinitrophenol is administered to a rat, its respiratory rate increases and its body temperature rises, as might be expected if the energy of oxidation is not conserved as ATP energy but rather is lost as heat.

5–6 THE ENERGY BALANCE SHEET FOR GLUCOSE OXIDATION

We are now in a position to sum up all the ATP molecules formed during the complete aerobic oxidation of glucose. First, we must remember that glucose is converted not to lactate but to pyruvate when the cell is aerobic. We therefore have the following equation for the glycolytic phase of glucose breakdown under aerobic conditions

$$\text{glucose} + 2\text{NAD}_{ox} + 2\text{ADP} + 2\text{P}_i \longrightarrow 2 \text{ pyruvate} + 2\text{NAD}_{red} + 2\text{H}_2\text{O} + 2\text{ATP} \qquad (5\text{–}10)$$

Now we can write the equation for the oxidation of the two molecules of NAD_{red} formed in the glycolytic sequence; this oxidation ultimately proceeds via the respiratory chain and is accompanied by phosphorylation of ADP. Although we would expect three molecules of ATP to be generated in this process, actually only two are formed in most cells on oxidation of NAD_{red} generated by glycolysis. This is because the NAD_{red} formed during glycolysis is *outside* the mitochondria and the pathway taken for entry of its electrons into the mitochondria is such as to yield only two molecules of ATP per pair of electrons. Our equation is thus

$$2\text{NAD}_{red} + 4\text{P}_i + 4\text{ADP} + \text{O}_2 \longrightarrow 2\text{NAD}_{ox} + 6\text{H}_2\text{O} + 4\text{ATP} \qquad (5\text{–}11)$$

Next we can write the equations for the oxidation of two molecules of pyruvate to acetyl CoA and the oxidation of the two molecules of NAD_{red} so formed; these reactions take place *within* the mitochondria

$$2 \text{ pyruvate} + 2\text{NAD}_{ox} \xrightarrow{\text{CoA}} 2 \text{ acetyl CoA} + 2\text{CO}_2 + 2\text{NAD}_{red} \qquad (5\text{–}12)$$

$$2\text{NAD}_{red} + 6\text{P}_i + 6\text{ADP} + \text{O}_2 \longrightarrow 2\text{NAD}_{ox} + 8\text{H}_2\text{O} + 6\text{ATP} \qquad (5\text{–}13)$$

Now we can write the overall equation for the oxidation of two molecules of acetate by two revolutions of the tricarboxylic acid cycle; in each revolution four pairs of electrons pass down the chain to oxygen and, on the average,* each pair causes formation of three molecules of ATP

* Actually, it is known that electron transport from succinate to oxygen causes only two phosphorylations, whereas electron transport from α-ketoglutarate to oxygen causes four. Oxidation of isocitrate and of malate cause three, to yield an average of three phosphorylations of ADP per pair of electrons for the entire cycle.

$$2 \text{ acetate} + 24P_i + 24ADP + 4O_2 \longrightarrow 4CO_2 + 28H_2O + 24ATP \quad (5\text{–}14)$$

Now let us add up all these equations (Eqs. 5–10 through 5–14) and strike out terms appearing on both sides. We will obtain

$$\text{glucose} + 36P_i + 36ADP + 6O_2 \longrightarrow 6CO_2 + 42H_2O + 36ATP \quad (5\text{–}15)$$

This is the overall equation for the complete oxidation of one molecule of glucose in aerobic cells; coupled to this process is the formation of thirty-six molecules of ATP from ADP and P_i. Since the standard free energy of oxidation of glucose is −686 kcal, and thirty-six molecules of ATP are formed, each requiring a minimum input of 7.3 kcal, the approximate efficiency of energy conservation is thus

$$\left(\frac{36 \times 7.3}{686}\right)100 = 38\%$$

However, it is very likely that this value is only a minimum figure. If we allow for the concentrations of ADP, ATP, and phosphate actually occurring in the cell, the efficiency of energy recovery during glucose oxidation in the intact cell is considerably higher, probably over 60 per cent.

5–7 REGULATION OF THE RATE OF RESPIRATION

We have already seen that the relative concentrations of ADP and ATP in the cell regulate the rate of glycolysis, through their capacity to act as positive and negative modulators, respectively, for the activity of the regulatory enzyme phosphofructokinase. Now we must ask how respiration, the aerobic phase of glucose degradation, is regulated. Moreover, we must also ask another vital question: How are the rates of glycolysis and respiration regulated with respect to each other, so that only enough pyruvic acid is produced by glycolysis to keep pace with the rate of utilization in the tricarboxylic acid cycle?

Just as the glycolytic sequence has a controlled reaction, or "pacemaker," which is catalyzed by phosphofructokinase, the tricarboxylic acid cycle also has a rate-limiting reaction catalyzed by a regulatory enzyme (Fig. 5–7). This is the dehydrogenation of isocitric acid to α-ketoglutaric acid, catalyzed by isocitric dehydrogenase

$$\text{isocitric acid} + NAD_{ox} \longrightarrow \alpha\text{-ketoglutaric acid} + CO_2 + NAD_{red}$$

The rate of the isocitric dehydrogenase reaction is largely modulated by ADP and ATP, neither of which is a substrate or product of the reaction catalyzed

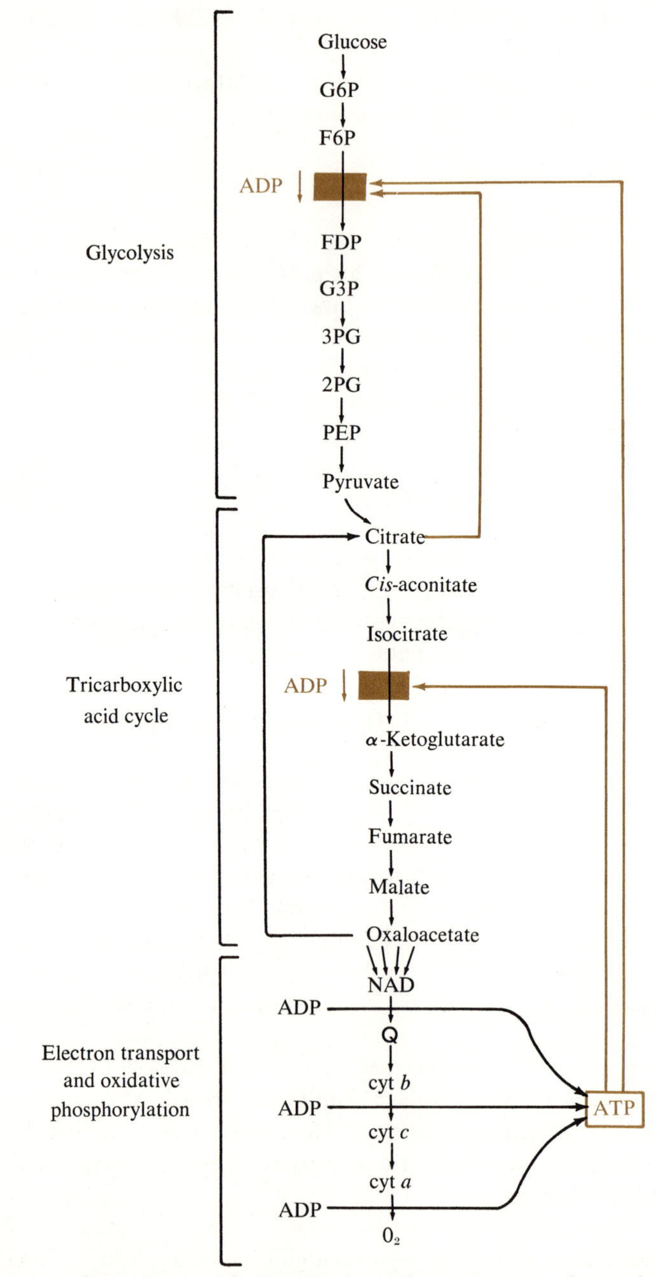

Figure 5–7. Regulation of glycolysis and respiration through feedback inhibition by ATP and citrate and positive modulation by ADP.

by the enzyme. ADP is a positive modulator and accelerates the isocitrate dehydrogenase reaction; ATP is a negative modulator and inhibits it. Thus whenever the concentration of ATP in the cell is relatively high and that of ADP is low, both the glycolytic sequence and the tricarboxylic acid cycle are throttled down. Whenever the ATP concentration is low and that of ADP is high, both are speeded up.

Moreover, Fig. 5–7 shows how the rate of glycolysis and the rate of the tricarboxylic acid cycle are integrated with each other. We have seen that phosphofructokinase is modulated by ATP and ADP. But it has another modulator, namely citric acid, an intermediate of the tricarboxylic acid cycle, which functions as a negative modulator. Whenever the citrate concentration in the cell exceeds a certain limit, indicating that citrate is being overproduced, then the rate of the phosphofructokinase reaction is depressed, because citrate, like ATP, causes isocitric dehydrogenase to be converted into its inactive form.

Spectroscopic observations of intact, respiring mitochondria have shown that the electron carrier molecules of the respiratory chain participate in an exquisitely balanced dynamic steady state, as is shown in Fig. 5–8, which also responds to the ratio of ADP to ATP. Normally the electron carriers nearest the electron-donating substrate are relatively reduced (i.e., filled with electrons) and those nearest oxygen are relatively oxidized, or empty of electrons; the intermediate carriers exist in a gradient, as is shown by the hydraulic model. This steady-state condition of the respiratory carriers is very delicately "tuned" to the concentration and ratio of ADP and ATP, in such a way that electron transport speeds up and the carriers become more oxidized when excess ADP is present. Conversely, when ADP is low in concentration, electron transport slows down and the carriers become more reduced.

Thus the glycolytic sequence, the tricarboxylic acid cycle, and the respiratory chain have self-adjusting and self-regulating features, so that the rate of the overall process of respiration is geared to the needs of the cell for ATP.

5–8 MITOCHONDRIA AND THEIR MOLECULAR ORGANIZATION

In Section 1–9, it was pointed out that there is a division of labor within the cell so that each of the various energy-yielding and energy-requiring processes occurs in one or another intracellular compartment or *organelle* of the cell. The enzymes of the Krebs tricarboxylic acid cycle and of the respiratory chain, as well as those responsible for making ATP during oxidative phosphorylation, have been found to be entirely located in the mitochondria, membrane-surrounded structural elements in the cytoplasm (Section 1–9). Let us now examine these structures in more detail because in the mitochondrion we have today perhaps the most advanced picture yet developed of the

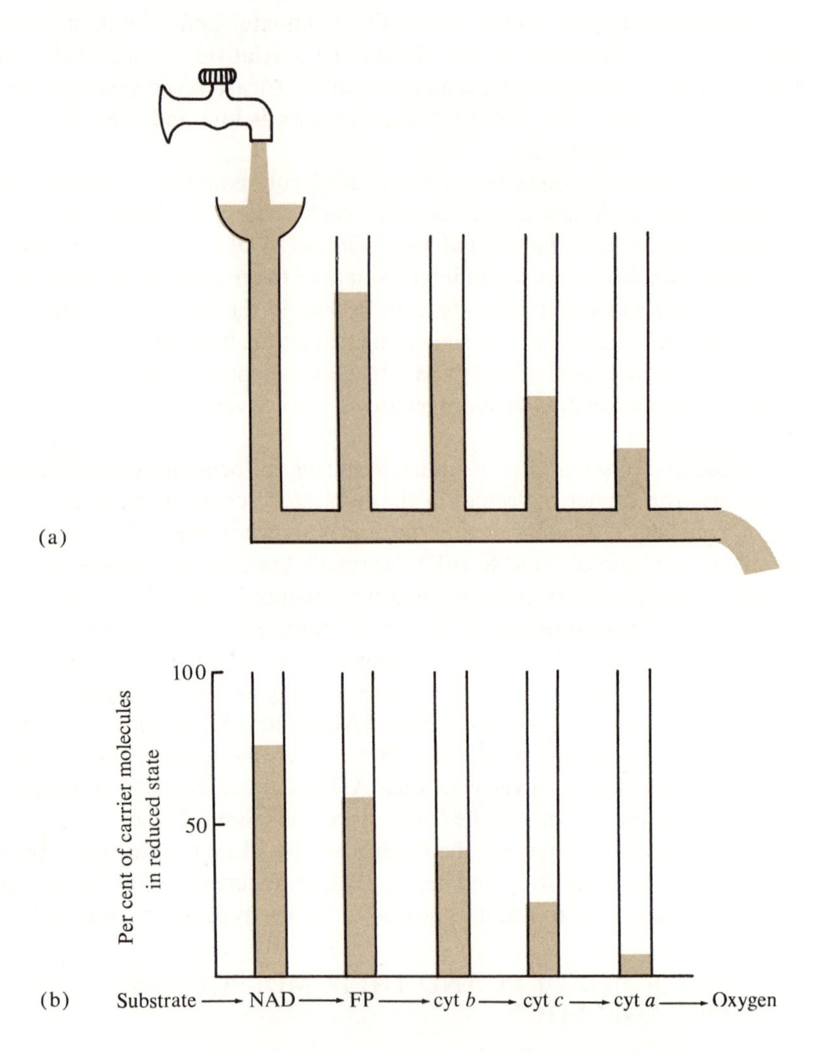

Figure 5–8. (a) Hydraulic model of the steady state of the respiratory chain. If the inflow of water is adjusted carefully, the levels of water in the open tubes will remain nearly constant. (b) The steady state of the respiratory chains in respiring mitochondria.

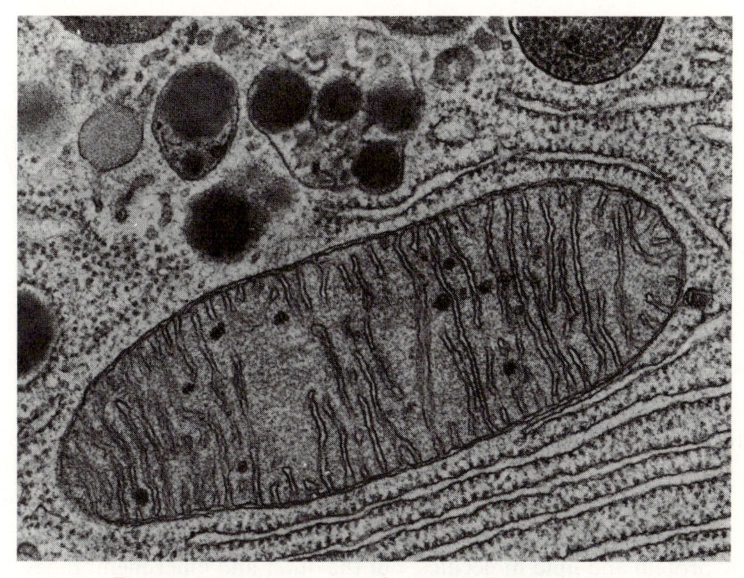

Figure 5–9. Electron micrograph of a mitochondrion in acinar cell of bat pancreas (courtesy of Dr. Keith Porter, University of Colorado).

molecular relationship between the ultrastructure of a cell component and its biochemical function.

Mitochondria may have many shapes from one type of cell to another. In the liver cell or muscle cell, however, they are approximate ellipsoids which are about 2 μ long and perhaps 1 μ in thickness. All aerobic eukaryotic cells possess mitochondria in their cytoplasm, but bacteria contain none. The number of mitochondria in a eukaryotic cell varies widely, from just a few dozen in some yeast cells to as many as 200,000 in very large egg cells of vertebrates. In the liver cell there are about 1000. Mitochondria are often strategically located near a source of fuel, such as lipid droplets, or near a site of ATP utilization, such as the myofibrils of muscle tissue.

The detailed structure of the mitochondrion has been revealed by the electron microscope. Figure 5–9 shows an electron micrograph of a single mitochondrion in a muscle cell, and Fig. 5–10 a schematic representation to show its structural details. The mitochondrion has two continuous membrane systems or "skins." The outer one is smooth, whereas the inner one is puckered inward in a series of folds, which are called *cristae*. In mitochondria of tissues with high rates of respiration, the cristae are very numerous and may nearly fill the entire inner compartment. The inner membrane, therefore, has

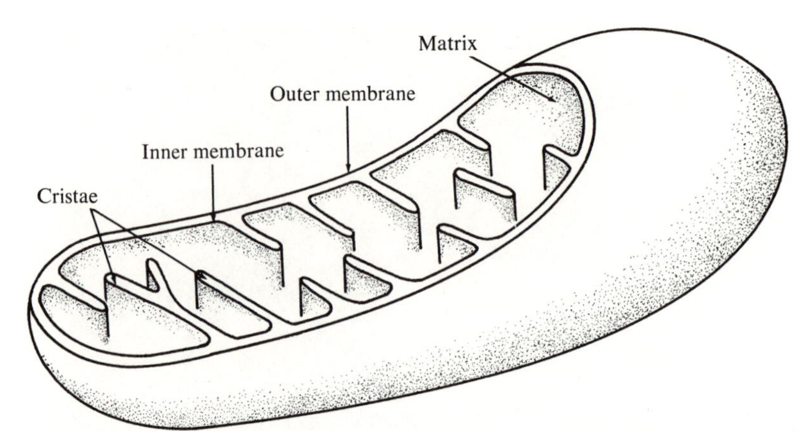

Figure 5–10. Schematic representation of mitochondrial structure.

a much larger surface area than the outer one. Within the inner compartment there is a semifluid material called the *matrix.* Both membranes contain specific protein and lipid molecules, but the outer and inner membranes differ somewhat in their chemical composition and properties. For example, the outer membrane is freely permeable to most small molecules, whereas the inner membrane is selectively permeable. It allows water, phosphate, ADP, ATP, and substrates such as pyruvate to pass freely, but K^+, Na^+, sucrose, and certain other polar molecules do not pass readily.

When cells or tissues are carefully ground to disrupt the cell membrane, the mitochondria are released and can be isolated by differential centrifugation. When such isolated mitochondria are suspended in a buffered medium at pH 7.0 containing only pyruvate, oxaloacetate, phosphate, and ADP, they oxidize the pyruvate very rapidly at the expense of molecular oxygen; simultaneously ADP is phosphorylated to ATP. Mitochondria thus behave as self-sufficient units; they contain all the enzymes and coenzymes required to carry out the complete oxidation of pyruvate and fatty acids via the tricarboxylic acid cycle, as well as the coupled phosphorylations.

Great strides have been made in studying the structure of the mitochondria in relation to their function as power plants. For example, it has been found that the enzymes of the tricarboxylic acid cycle are located in the matrix within the inner compartment of the mitochondrion. In contrast, the enzymes of the respiratory chain are exclusively located in the inner membrane. Furthermore, it appears probable that the cytochromes are actually located next to each other in the membrane in the exact sequence in which they interact. A set or cluster of cytochromes, together with the flavoproteins that feed into the cytochrome system, is called a *respiratory assembly* (Fig. 5–11). Such clusters of cytochromes are located in the inner membrane at regular intervals, about 220 Å apart. A single liver mitochondrion contains about 15,000 such

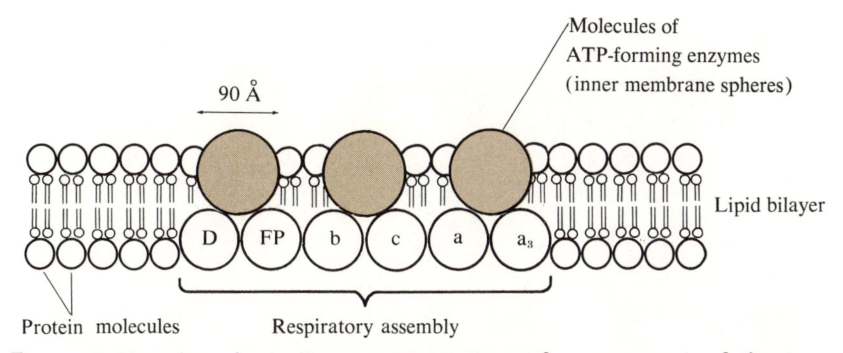

Figure 5–11. A schematic representation of a segment of the inner mitochondrial membrane, showing respiratory assemblies and inner membrane spheres.

clusters. The inner mitochondrial membrane is therefore not simply an inert sheet or skin, but a complex molecular fabric of lipid and protein molecules in which the cytochromes and other electron carriers are fixed.

The ATP-forming enzyme has a relative high molecular weight. Special staining methods reveal that its molcules project from the inner surface of the inner membrane (Fig. 5–12). When these knob-like particles are removed from the inner mitochondrial membrane, the latter loses its capacity to phosphorylate ADP during electron transport. However, this capacity can be restored again on adding the ATP-forming enzyme back to preparations of the inner membrane lacking the knobs.

5–9 THE MECHANISM OF OXIDATIVE PHOSPHORYLATION

Despite years of effort the mechanism by which ATP is generated during electron transport is still unknown. However, it is clear that the inner mitochondrial membrane is an essential element in the mechanism by which the energy of electron transport is recovered as phosphate bond energy. If this membrane is damaged or disrupted, then oxidative phosphorylation of ADP no longer occurs, although electron transport may still take place. When electron transport takes place in normal intact mitochondria, the inner membrane undergoes characteristic changes in the manner in which it is folded to form the cristae. Moreover, during electron transport H^+ ions may be pumped out of mitochondria. These observations have suggested three possible mechanisms by which electron transport energy is converted into phosphate bond energy. The *chemical coupling hypothesis* (Fig. 5–13) holds that oxidative phosphorylation is catalyzed by a sequence of enzymes acting consecutively

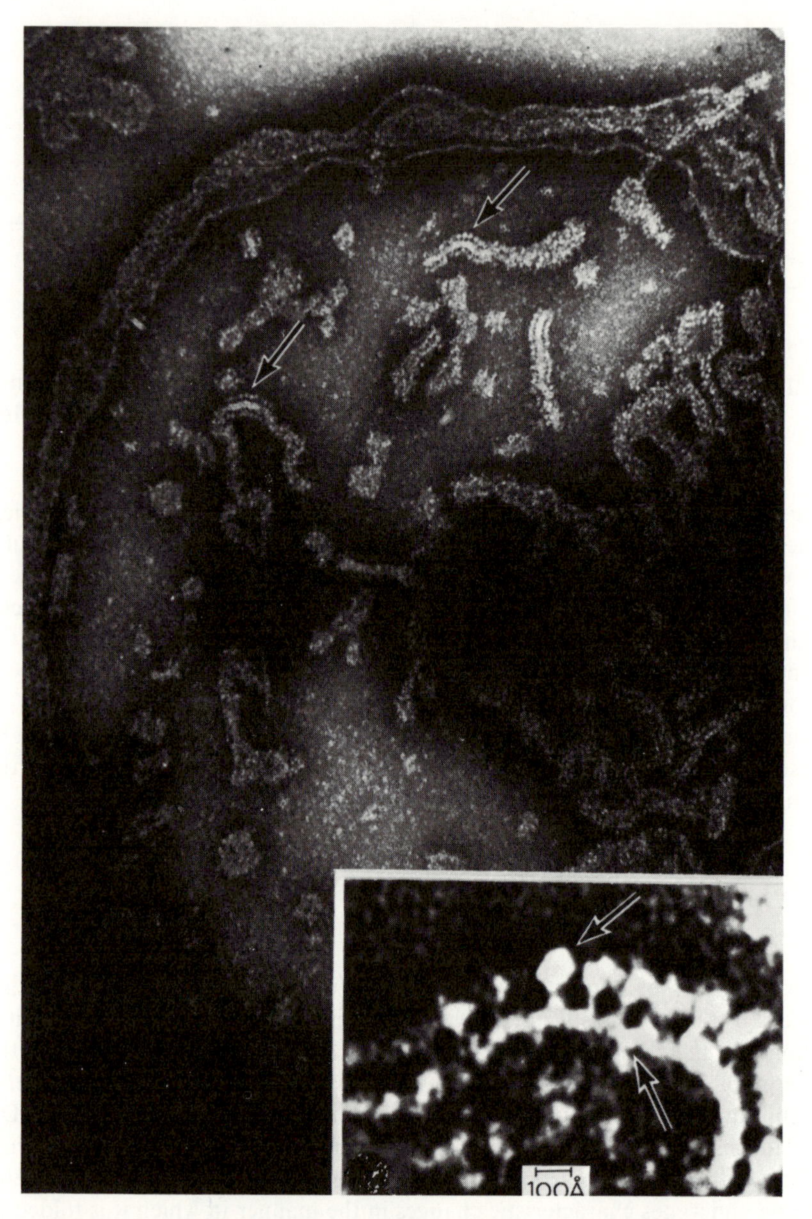

Figure 5–12. An electron micrograph of the inner membrane of a heart mitochondrion, showing the knoblike particles of the ATP-synthesizing enzyme molecules (arrows). The insert at the lower right represents a higher magnification. (courtesy of Dr. H. Fernández-Morán, University of Chicago.)

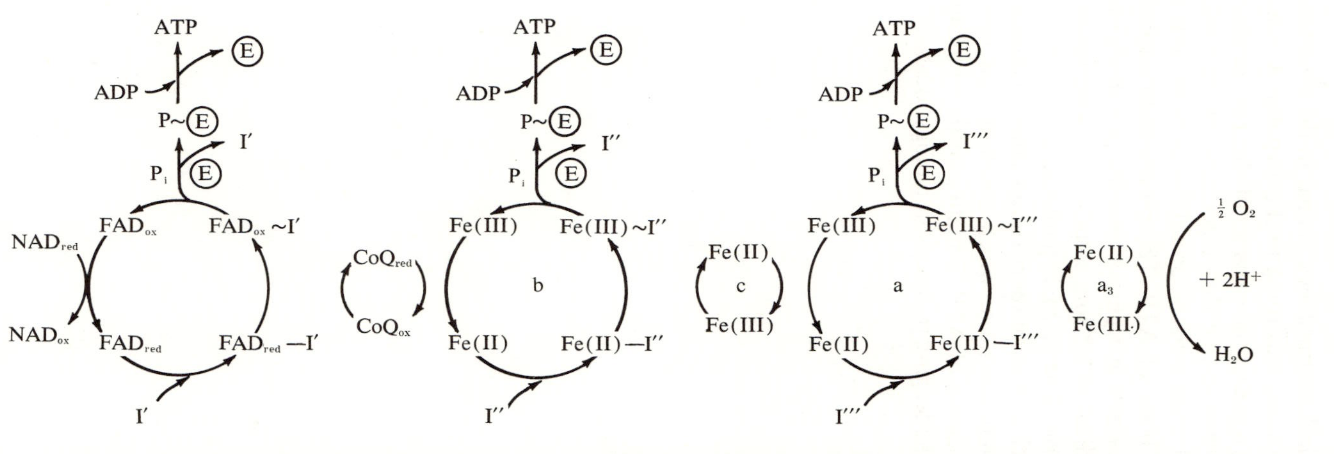

Figure 5–13. A chemical coupling hypothesis for oxidative phosphorylation. The components I′, I″, and I‴ are coupling factors and E is a phosphate-transferring enzyme. ATP is formed in a sequence of reactions with common intermediates at the expense of the high-energy intermediates $FAD_{ox} \sim I'$, $Fe(III) \sim I''$, and $Fe(III) \sim I'''$, which are postulated to be generated by electron transport.

through common chemical intermediates, similar to the consecutive reactions of glycolysis. The membrane, according to this hypothesis, must supply the framework for bringing together the enzymes of electron transport and phosphorylation in order to ensure that they react consecutively, with formation of high-energy chemical intermediates, postulated to be the precursors of the high-energy phosphate bond of ATP. The *chemiosmotic hypothesis*, on the other hand, holds that electron transport pumps protons across the membrane, and that the gradient of H^+ ions so produced is the immediate driving force for ATP formation. The third hypothesis, called the *conformational hypothesis*, holds that electron transport causes conformational changes in the membrane of the mitochondria, which are converted into phosphate-bond energy. The mechanism of ATP formation in the mitochondria is one of the most lively and challenging problems in biology today.

6

PHOTOSYNTHESIS AND THE CHLOROPLAST

Now we come to consider how light energy, the ultimate source of all biological energy, is absorbed by photosynthetic cells and converted into chemical energy, which in turn is then used to reduce carbon dioxide to form glucose. Because photosynthesis is the first stage in the flow of energy through the biosphere, it may appear illogical for us to have discussed glycolysis and respiration before photosynthesis. However, photosynthesis is a more complex and less well understood process than respiration. Moreover, it was necessary to develop some of the bioenergetic principles underlying the glycolytic system, the tricarboxylic acid cycle, and electron transport, because we will soon find that these principles are also involved in the action of the photosynthetic systems of plant cells.

6–1 THE EQUATION OF PHOTOSYNTHESIS

The equation for the formation of glucose and molecular oxygen from CO_2 and H_2O, which requires the input of free energy, may be written as follows

$$6CO_2 + 6H_2O \longrightarrow C_6H_{12}O_2 + 6O_2 \qquad \Delta G^{0\prime} = +686 \text{ kcal}$$

During photosynthesis in plants the free energy required for this reaction is furnished by photons or light quanta and the equation becomes

$$6CO_2 + 6H_2O + n\,h\nu \longrightarrow C_6H_{12}O_6 + 6O_2$$

Although this equation describes the photosynthetic process in higher plants, there are other kinds of photosynthetic organisms that carry out variations on this basic equation. For example, the group of pigmented photosynthetic bacteria called *purple bacteria* do not produce oxygen during photosynthesis. Instead of using H_2O as a hydrogen donor, they use H_2S, and instead of producing oxygen, they produce elementary sulfur, according to the following equation

$$6CO_2 + 12H_2S + n\,h\nu \longrightarrow C_6H_{12}O_6 + 12S + 6H_2O$$

Still other photosynthetic bacteria use certain organic molecules as hydrogen donors, as in the equation

$$6CO_2 + \underset{\text{isopropanol}}{12CH_3\text{—}CHOH\text{—}CH_3} + n\,h\nu \longrightarrow$$

$$C_6H_{12}O_6 + \underset{\text{acetone}}{12CH_3\text{—}\underset{\underset{\displaystyle O}{\|}}{C}\text{—}CH_3} + 6H_2O$$

An important conclusion has been drawn from such comparative studies of the photosynthetic equation in different kinds of cells. The CO_2 that is "fixed" to yield glucose gains hydrogen atoms at the expense of a hydrogen donor, which may be either H_2O, or H_2S, or some other reducing agent such as isopropanol. In effect, the photosynthetic process causes hydrogen atoms to be transferred from a hydrogen donor, such as H_2O or H_2S, to carbon dioxide, which becomes reduced to glucose. The dehydrogenated form of the hydrogen donor, such as oxygen or sulfur, is the other major product, as is shown in the generalized equation

$$6CO_2 + 12H_2X + n\,h\nu \longrightarrow C_6H_{12}O_6 + 12X + 6H_2O \qquad (6\text{–}1)$$

where H_2X is the hydrogen donor and X its dehydrogenated form. Of course, as in all biological oxidoreduction reactions, the ultimate entities transferred are really electrons. Thus H_2X is the electron donor and CO_2 is the electron acceptor in photosynthesis. This generalized equation of photosynthesis was proposed by van Niel, a pioneer in the study of photosynthesis and the comparative aspects of metabolism.

Equation 6–1 also emphasizes that plant photosynthesis is the reverse of respiration. In respiration, hydrogen atoms are released from substrates and the equivalent electrons combine with oxygen at the end of the respiratory chain; a large amount of energy is thus released as H_2O is formed. In plant photosynthesis, electrons are removed from the H_2O molecule and transferred "uphill" to CO_2, a process that requires input of energy. Equation 6–1 also suggests that the oxygen gas formed in photosynthesis must arise from the H_2O molecule and not the CO_2; this view has been supported by experiments in which water labeled with the ^{18}O isotope was shown to give rise to molecular oxygen labeled with ^{18}O during photosynthesis.

6–2 THE "LIGHT" AND "DARK" REACTIONS

The term photosynthesis usually refers to the total process by which glucose is formed from CO_2 and H_2O at the expense of solar energy. However, light is directly required only in the first stages of photosynthesis, in which the captured light energy is converted into chemical energy; these early reactions are therefore called the *light reactions*. The remaining reactions by which glucose is formed from CO_2 at the expense of chemical energy can proceed in the absence of light, and they are called the *dark reactions*. The light reactions are unique to photosynthesizing cells, whereas most of the dark reactions, by which the carbon skeleton of glucose is built from CO_2 and H_2O, also occur in many heterotrophic cells.

6–3 LIGHT QUANTA

Visible light is a form of electromagnetic radiation; its wavelength falls in the zone between about 400 nm and 700 nm. Just below this zone we have the ultraviolet and just above, the infrared region (Fig. 6–1).

In many of its properties, light behaves as though it consists of waves. For example, light rays can be bent or diffracted and can also be spread so that the edge of a shadow is not perfectly sharp. To the contrary, certain other properties of light are more consistent with the view that it is corpuscular. For example, when light beams are directed at surfaces of certain substances such as selenium, electrons are ejected from its atoms. The greater the intensity of the light, the larger the number of electrons ejected. This phenomenon, the *photoelectric effect,* is made use of in the familiar photocell. As a result of these two types of observations, light is now thought to consist of waves of particles known as *photons*. It was Einstein who first proposed that photons are actually units of energy, or light quanta.

The energy equivalent of a quantum of light is not fixed; rather it depends on the wavelength (Fig. 6–2). Photons of long wavelength, at the red end of

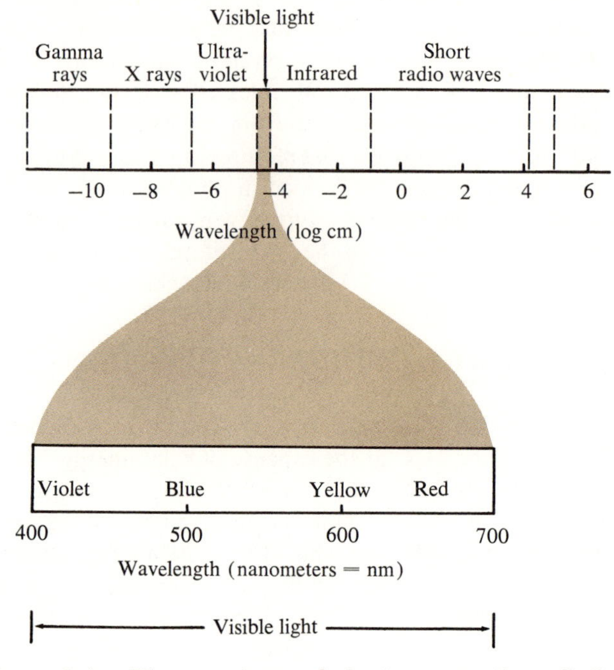

Figure 6–1. The spectrum of electromagnetic radiation.

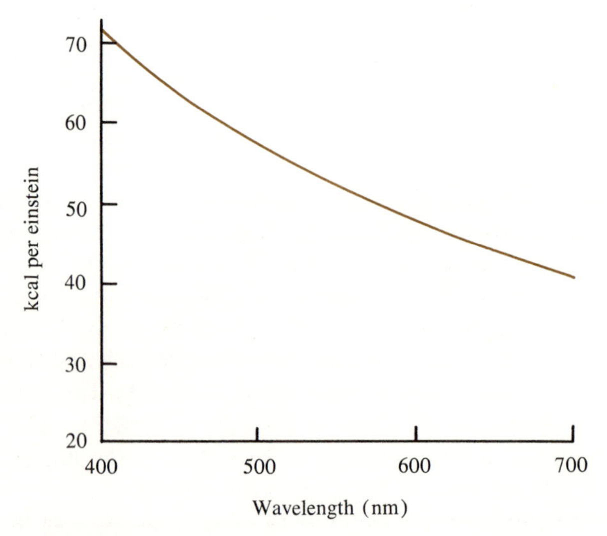

Figure 6–2. The energy content of photons.

the spectrum, have the least energy, whereas photons of short wavelength, at the blue end of the spectrum, have the greatest energy. The fundamental equation relating the energy content of photons to their wavelength or frequency is

$$E = h\nu = h\frac{c}{\lambda}$$

in which E is the energy of a light quantum, ν is the frequency in vibrations per second, h is Planck's constant, c is the velocity of light, and λ the wavelength.

The molar unit for expressing the amount of light energy is the *einstein.* This is the number of photons in one "mole" of light; it has the same numerical value as *Avogadro's number*, which is the number of molecules in a gram molecular weight of a substance, namely 6.023×10^{23}.

6–4 EXCITATION OF MOLECULES BY LIGHT

The ability to absorb light varies greatly from one substance to another. Water evidently absorbs very little visible light of any wavelength and therefore appears colorless. On the other hand a solution of dye molecules strongly absorbs light, but only at certain wavelengths, and for this reason the transmitted light has a characteristic color. If we plot the ability of a substance to absorb light against the wavelength of the light, we obtain an *absorption spectrum*. As an example Fig. 6–3 shows the absorption spectrum of chlorophyll *a*. We see that chlorophyll strongly absorbs light in the regions 400–450 nm (violet) and 640–660 nm (red). As a result the light transmitted by a chlorophyll solution, in the region 450–640 nm, appears green.

Light absorbed by an atom or molecule is actually absorbed by certain of its electrons. Electrons are arranged in different orbitals around the nucleus of each atom. Those electrons nearest the nucleus have relatively low energy, and those electrons farthest from the nucleus have a higher energy. To move an electron from an inner to an outer position requires input of energy because a negatively charged particle is being moved away from the positively charged nucleus. When photons strike an atom or molecule that can absorb light, an inner electron may absorb the photon and thus gain its energy, which then becomes sufficient to move the electron farther away from the nucleus to an outer position with a higher energy level. When this happens, the atom is spoken of as being in the *excited state*. Only certain wavelengths of light can excite specific atoms because the sensitive electron must absorb a photon having an energy content that is at least equal to the energy difference between the inner orbital and the available outer orbital to which the electron can be "boosted." A photon having less energy cannot excite the atom. The energy in a photon thus is used on an "all or none" basis, hence the term *quantum*.

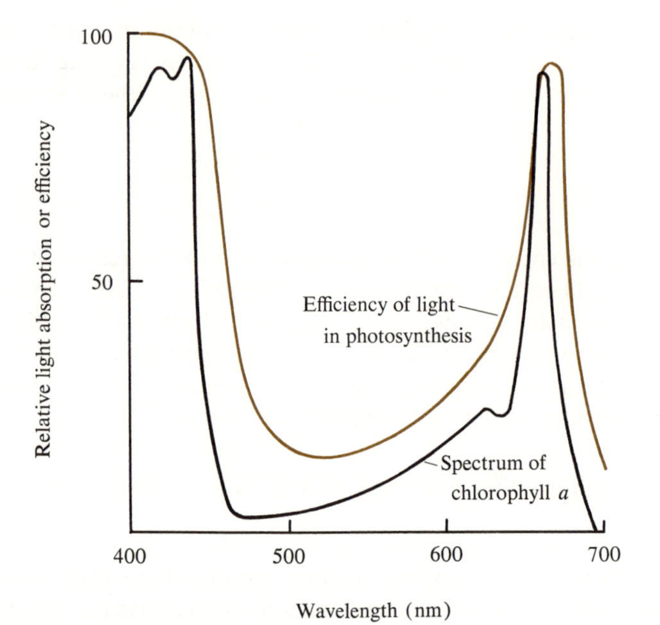

Figure 6–3. Absorption spectrum of chlorophyll *a*. The line in color shows the efficiency spectrum of light in photosynthesis.

Excited atoms or molecules are very unstable, since the high-energy electrons tend to return to their original low-energy orbitals again. When this happens the molecule is said to return to the *ground* state. Such a return of a high-energy electron to its original orbital obviously must proceed with release of the light energy that was originally absorbed from the photon. Some of this energy may appear as heat, and some of it may reappear as light. Such an emission of light from an excited molecule, as it returns to its ground state, is called *fluorescence*. However, when light quanta are absorbed by electrons of some photosensitive atoms, such as those of the selenium in a photocell, the electrons acquire such a high energy that they completely escape from the selenium atoms and may be tapped off by means of a wire. In this way absorbed light sets up an electrical current (i.e., an electron flow) in a photocell. We shall presently see that a process similar to that occurring in a photocell is involved in photosynthesis.

6–5 CHLOROPHYLL

Since only absorbed light can excite molecules and thus deliver its energy, it must be the pigments of photosynthetic cells that act as absorbers of visible light. Our attention is at once focused on chlorophyll. The leaves of higher plants actually contain two kinds of chlorophyll, which differ only slightly in structure and absorption spectrum, *chlorophyll a* and *chlorophyll b*. As

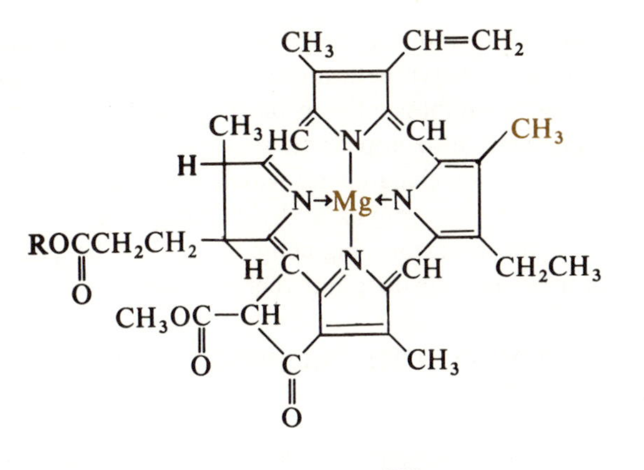

O
‖
—C—H
Formyl group

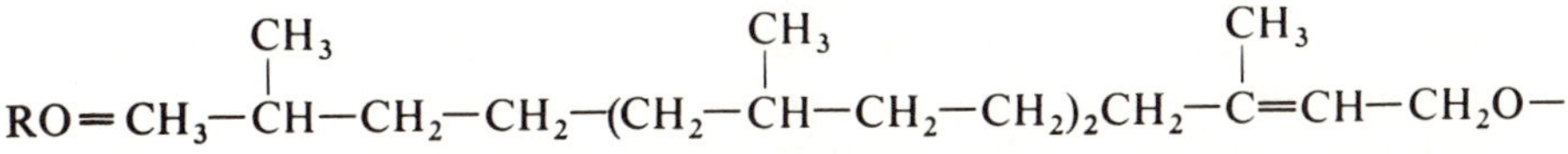

Phytyl group

Figure 6–4. Structure of chlorophyll *a*. The long chain alcohol phytol is esterified to the carboxyl group as shown. In chlorophyll *b* the formyl group replaces the methyl group in color.

we shall see, each plays a specific role in the mechanism of photosynthesis in plants. The molecular structure of chlorophyll *a* is shown in Fig. 6–4. It consists of a nearly flat arrangement of four pyrrole rings grouped around a central atom of Mg^{2+}. Also noteworthy is the long hydrocarbon side chain of *phytol*, which lends this molecule solubility in lipids or fats. The structure of chlorophyll *b* is only slightly different; it contains a —CHO group instead of the methyl group at the position shown. The absorption spectrum of chlorophyll *a* is shown in Fig. 6–3. Chlorophyll has been concluded to be the main light-capturing molecule in most higher green plants because there is a close correspondence between the measured efficiency of different wavelengths of light in promoting photosynthesis and the absorption of light by chlorophyll, as is shown in Fig. 6–3.

Chlorophyll also occurs in photosynthetic bacteria, which we may recall do not produce molecular oxygen. However, these organisms contain only one chlorophyll, *bacteriochlorophyll*, in contrast to oxygen-evolving photosynthetic organisms, which always contain two types of chlorophyll. For this reason it has been concluded that one of the two chlorophylls of higher plants, presumably chlorophyll *b*, is specifically required for the evolution of oxygen.

Many photosynthetic cells contain, in addition to chlorophyll, other light-absorbing pigments, known as *accessory pigments*. These include the *carotenes* which are yellow, brown, or red compounds, and the *phycocyanins* and *phycoerythrins*, which are blue and red respectively. These pigments are especially prominent in some photosynthetic algae, which may be red, brown, green, or blue in color, and in photosynthetic bacteria, some of which are purple. The accessory pigments also appear to serve as absorbers of light energy; they absorb light at wavelengths at which chlorophylls are not effective and thus supplement chlorophylls as light "traps." However, the light energy absorbed by the accessory pigments must first be passed to chlorophyll before it can be used to do photochemical work.

6–6 THE EXCITATION OF CHLOROPHYLL AND PHOTOREDUCTION

But how can light energy absorbed by chlorophyll be converted into chemical energy? Pure chlorophyll isolated from leaves goes into an excited state when it is illuminated. However, no chemical work is done by excited chlorophyll molecules in the test tube because they simply return to their ground state and lose the absorbed light energy as fluorescence and heat (Fig. 6–5). But chlorophyll molecules in intact leaf cells behave quite differently on illumination. They do not lose their energy of excitation by simple fluorescence, but they behave much like the photoelectric cell we described above. When intact plant cells are illuminated, high-energy electrons leave the excited chlorophyll molecule entirely and are led away from it, not by a wire, as in a photocell, but by a chain of electron-carrier enzymes very similar to those that participate in electron transport in the mitochondria during respiration. We have already seen from our analysis of electron transport in

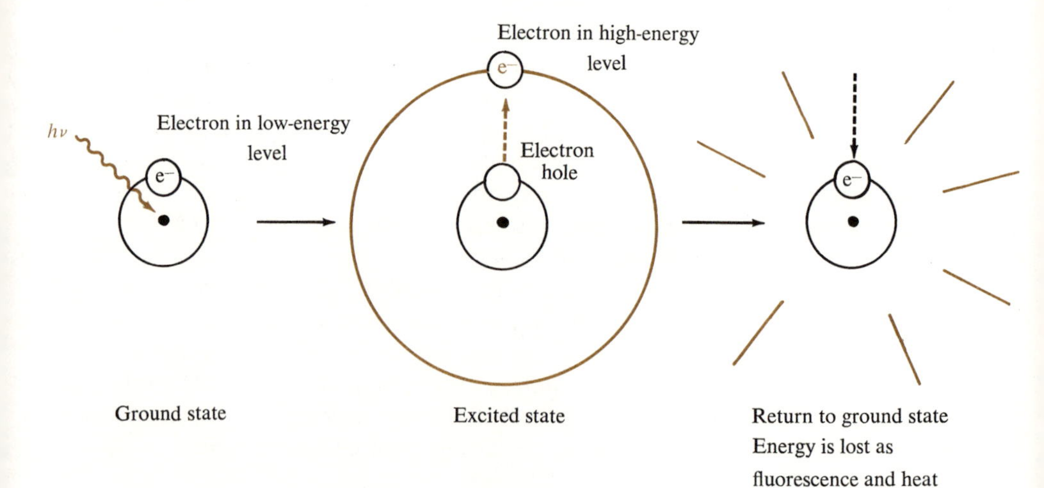

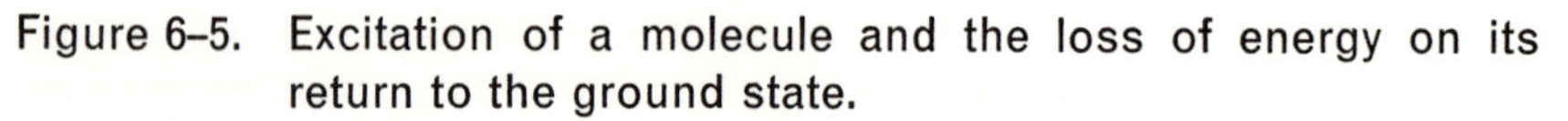
Figure 6–5. Excitation of a molecule and the loss of energy on its return to the ground state.

mitochondria that electrons are carriers of energy and when they flow downhill from one electron-carrier to another they release their energy. Now let us examine how electrons flow from excited chlorophyll molecules.

We have seen (Eq. 6–1) that the ultimate acceptor of electrons in plant photosynthesis is carbon dioxide, which becomes reduced to form sugar. In the late 1930's, Hill made a discovery that ultimately led to the identification of a number of electron carriers that function to transport electrons away from excited chlorophyll. He found that isolated chloroplasts from spinach leaves could, when illuminated, reduce various artificial electron acceptors such as ferricyanide or certain dyes with simultaneous production of oxygen. This observation started a search for naturally occurring electron acceptors in chloroplasts. Ultimately it was found that the oxidized forms of NAD and a closely related electron acceptor, namely *nicotinamide adenine dinucleotide*

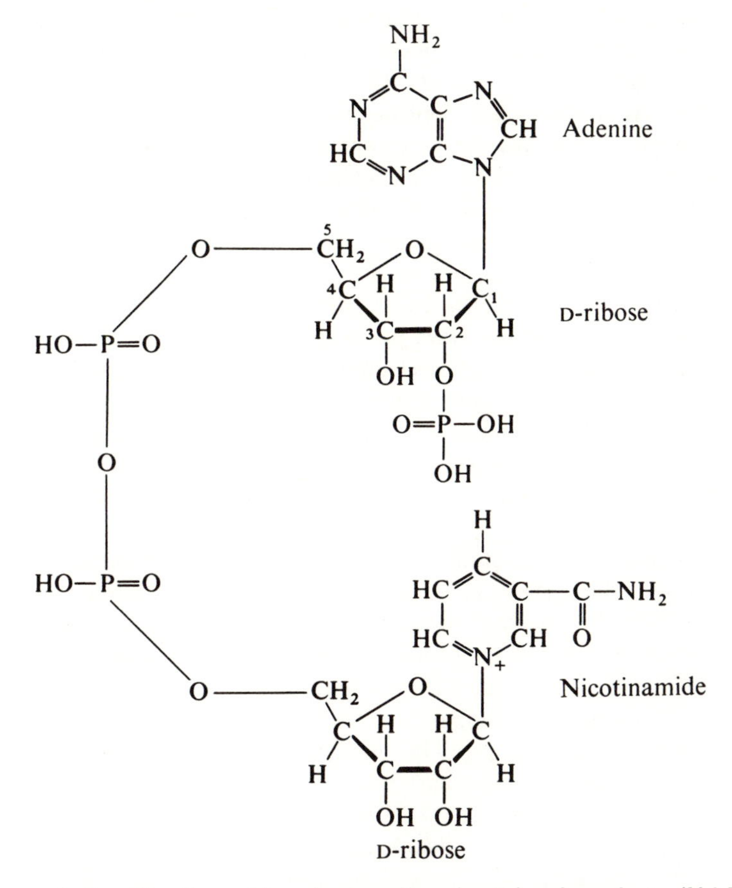

Figure 6–6. Nicotinamide adenine dinucleotide phosphate (NADP).

phosphate, abbreviated NADP, also serve to accept electrons when chloroplasts are illuminated. NADP differs from NAD only in having a third phosphate group substituted at the 2-hydroxyl group of the ribose next to the adenine ring (Fig. 6–6). Although NAD and NADP are structurally almost identical, and both are found in all living cells, whether photosynthetic or heterotrophic, they are not interchangeable and have quite different biochemical functions. NAD is the specific electron acceptor for those dehydrogenases that are normally concerned in passing electrons *toward* oxygen, as occurs during respiration. In contrast, NADP is specific for dehydrogenases that primarily function to provide electrons for the reduction of organic substrates, for example, the reduction of carbonyl groups to alcohols or of carbon-carbon double bonds to form single bonds. As we shall see, $NADP_{red}$ is the most important and most immediate reducing agent required for reduction of carbon dioxide to sugar in the dark reactions of photosynthesis.

But how is NADP reduced by excitation of chlorophyll? Much research has revealed that there is a chain of electron carriers leading from chlorophyll *a* molecules to NADP in chloroplasts. The constitution of this chain is shown in Fig. 6–7. When an electron is "boosted" to a high-energy level on excitation of chlorophyll *a*, it leaves the latter and passes to another pigment called P700, which in itself appears to be a specialized chlorophyll molecule. P700 acts as a "tap" from an assembly of light-absorbing molecules, which includes chlorophyll *a* and carotenoids. These "hot" electrons then are passed to a specialized electron-carrying protein that contains iron and sulfur, namely *ferredoxin.* This electron carrier differs from the iron-containing cytochrome pigments in that it does not have porphyrin groups; nevertheless, the iron of ferredoxin undergoes transitions from FeII to FeIII as it transfers electrons to $NADP_{ox}$ via a flavoprotein electron carrier, *ferredoxin-NADP oxidoreductase.* This light-induced flow of electrons from chlorophyll to NADP is called *noncyclic electron flow*; it continues until all available NADP is reduced. Noncyclic electron flow thus proceeds with the accumulation of a reduced product.

There is another type of light-induced electron transport, called *cyclic* electron flow, which takes place when chloroplasts are illuminated in the *absence* of $NADP_{ox}$ as electron acceptor. In cyclic electron flow electrons leave excited chlorophyll molecules on illumination, pass along a circular chain of electron carriers, and then *return* to the chlorophyll molecule again; no reduced product accumulates during cyclic electron flow. This circular chain of electron carriers probably includes some of the electron carriers used in noncyclic electron flow, particularly P700 and possibly also ferredoxin. In addition it also contains two cytochromes, cytochrome *b* and cytochrome *f*, which resemble but are not identical to the cytochromes functioning in electron transport mitochondria. The chain of electron carriers functioning in cyclic electron flow is shown in Fig. 6–8.

But the question now arises: If illumination of chloroplasts causes electrons

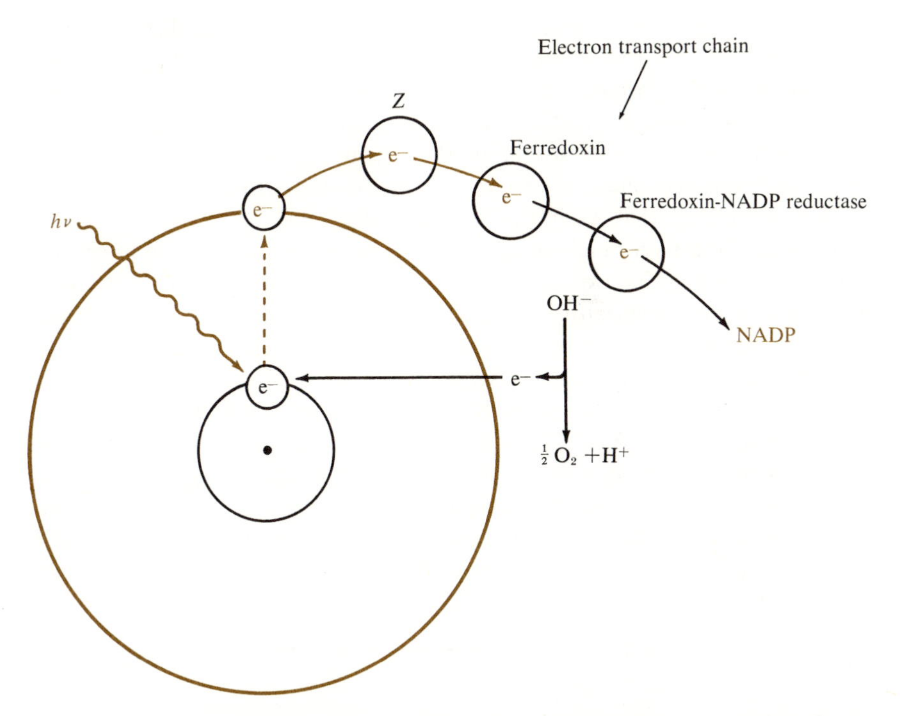

Figure 6–7. Noncyclic electron flow. The high-energy electrons from excited chlorophyll are used to reduce $NADP_{ox}$ to $NADP_{red}$. The electron holes left behind are filled with electrons from hydroxyl ions of water. The letter *Z* represents an unknown carrier thought to be the initial electron acceptor from P_{700}.

to flow out of the excited chlorophyll molecule, around the chain, and then back to chlorophyll again, thus restoring the normal number of electrons in the latter, what has been accomplished by cyclic electron flow? What can be the purpose of this process? Moreover, how can one even detect cyclic electron flow?

6–7 PHOTOPHOSPHORYLATION

Actually the occurrence of cyclic photoinduced electron flow in chloroplasts was first detected because of an effect it produces. In the early 1950's it was discovered that when isolated chloroplasts are illuminated in the presence of ADP and phosphate, ATP is formed at a high rate. The ATP formation was found to occur in the complete absence of organic substrates and of oxygen. Therefore, the observed phosphorylation was clearly not due to a flow of electrons from substrates to oxygen, as occurs in oxidative phosphorylation in mitochondria. However, the phosphorylation of ADP by illuminated chloroplasts was directly dependent on the intensity of light and the duration of illumination; the longer the chloroplasts were illuminated, the greater the amount of ATP formed. This type of phosphorylation was christened *photosynthetic phosphorylation*, or more simply, *photophosphorylation*. Because the only possible source of the energy required to make the ATP from ADP and phosphate was the radiant energy supplied to the chloroplasts, it was postulated that excitation of chlorophyll led to the generation of electrons having a very high energy level. As these electrons return to chlorophyll again, via a chain of carriers, they lose their energy, just as electrons passing down the respiratory chain of mitochondria lose their energy. It was also postulated that at one or more points along this closed chain of electron carriers there are enzymatic mechanisms, similar to those in mitochondria, which convert the oxidation-reduction energy of electron flow into the phosphate bond energy of ATP. Thus photoinduced cyclic electron flow has a real and important purpose, namely, to transform the light energy absorbed by the chlorophyll molecules in the chloroplast into phosphate bond energy. For this reason such phosphorylation is called specifically *cyclic photophosphorylation*. The overall equation for cyclic photophosphorylation may be written as

$$n\,h\nu + P_i + ADP \longrightarrow ATP + H_2O$$

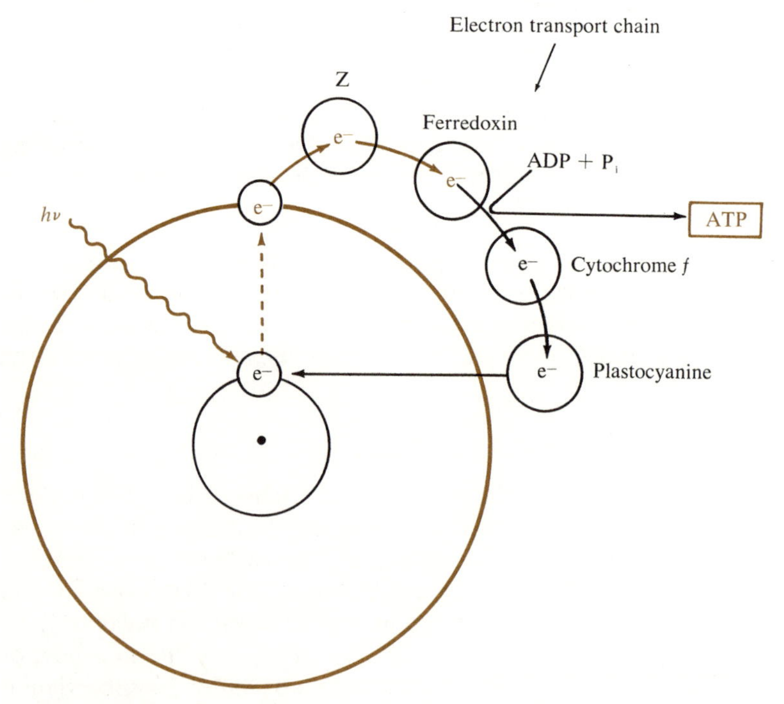

Figure 6–8. Cyclic electron flow and photophosphorylation.

Later it was found that photophosphorylation of ADP also occurs when chloroplasts are illuminated in the presence of $NADP_{ox}$, a process that we have already seen results in the reduction of $NADP_{ox}$ to $NADP_{red}$ and in the evolution of molecular oxygen. Photophosphorylation accompanying reduction of $NADP_{ox}$ is called *noncyclic photophosphorylation.* We may represent this process by the equation

$$n\,h\nu + 2P_i + 2ADP + H_2O + NADP_{ox} \longrightarrow NADP_{red} + \tfrac{1}{2}O_2 + 2ATP + 2H_2O$$

We have now seen that illumination of chloroplasts can lead to the formation of the two chemical agents that are necessary to carry out the biosynthesis of glucose from carbon dioxide, namely $NADP_{red}$ and ATP. But we do not yet have a clear picture of the "hook-up" or wiring diagram of the various enzymatic processes of photoinduced electron flow and photophosphorylation. How are cyclic and noncyclic electron flow related to each other? How is molecular oxygen formed? What is the relationship between chlorophyll *a* and chlorophyll *b*? By what mechanism does photophosphorylation take place? These questions cannot yet be fully answered, but let us now see how the information we have developed has been put together into a general hypothesis for the pattern of the light reactions of photosynthesis.

6–8 PHOTOSYSTEMS I AND II AND THEIR INTERRELATIONSHIPS

We have seen that higher plants, which evolve oxygen, contain two types of chlorophyll, whereas photosynthetic bacteria, which do not evolve oxygen, have but one type of chlorophyll. Chlorophyll *a* is the characteristic pigment of what is called *photosystem I* of higher plants; it is believed to play a primary role in photophosphorylation and in reduction of $NADP_{ox}$. Chlorophyll *b*, on the other hand, is the characteristic light-absorbing pigment of *photosystem II* in higher plants. Photosystem II is found only in oxygen-evolving photosynthetic cells and it is accordingly postulated to represent the system that generates oxygen from water. Photosystems I and II of higher plants can be excited independently by different wavelengths of light, since the absorption spectra of chlorophyll *a* and *b* differ significantly.

The interrelationships between photosystems I and II, and how they bring about photophosphorylation, photoreduction, and oxygen evolution, are currently the subjects of intensive research. One hypothesis, supported by most existing evidence, is shown diagrammatically in Fig. 6–9, which shows also the energy level of the electrons at various stages in the light reactions. It is postulated that photosystems I and II are connected to each other by a chain of electron carriers. Let us see, by tracing the pathway of electron flow, how

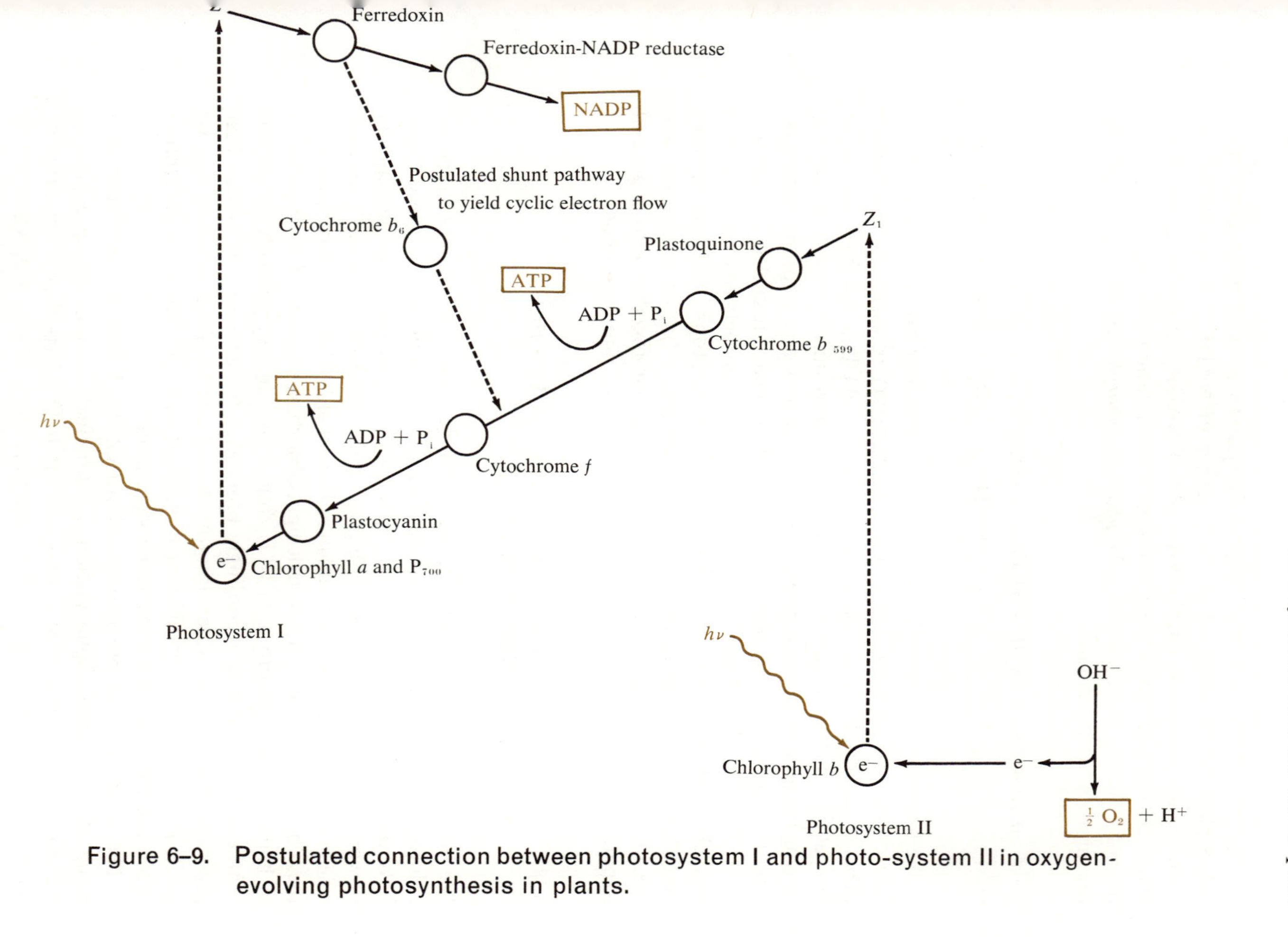

Figure 6–9. Postulated connection between photosystem I and photo-system II in oxygen-evolving photosynthesis in plants.

this connecting link makes possible coordination of photophosphorylation, photoreduction of $NADP_{ox}$, and oxygen evolution.

When chlorophyll *a* of photosystem I is illuminated, one or more electrons are boosted to a high-energy level and are tapped off to the first member of a chain of electron carriers. These electrons then pass via ferredoxin to $NADP_{ox}$, reducing the latter to $NADP_{red}$. But this process of photoreduction of $NADP_{ox}$ cannot continue long because each chlorophyll *a* molecule can give up only one or a very few electrons, leaving behind what are called electron "holes." Until these holes are filled with electrons chlorophyll *a* cannot return to the ground state.

The electrons required to return chlorophyll *a* of photosystem I to the ground state come from another electron transport chain that leads from photosystem II to photosystem I. When photosystem II is illuminated and chlorophyll *b* goes into the excited state, electrons of the latter are boosted to a high-energy level and pass down the connecting chain of carriers, losing energy as they go, until finally they enter chlorophyll *a* and bring it back to the ground state again, rendering photosystem I ready for another cycle of excitation and deexcitation. As the electrons flow downhill from excited photosystem II to the electron holes in chlorophyll *a*, some of their energy is utilized to cause photophosphorylation, the coupled synthesis of ATP from ADP and phosphate. Apparently two energy-conserving sites are involved in the connecting chain.

Although we have now accounted for the photoreduction of $NADP_{ox}$, the photophosphorylation, and the refilling of the electron holes left in chlorophyll *a* after its excitation, we must now ask: How are the electron holes in excited chlorophyll *b* refilled? As is seen in Fig. 6–9, the electron holes in chlorophyll *b* are believed to be filled with electrons arising from water, more specifically, from hydroxyl ions. Removal of electrons from hydroxyl ions causes ultimately the formation of molecular oxygen

$$H_2O \longrightarrow H^+ + OH^-$$

$$OH^- \longrightarrow \tfrac{1}{2}O_2 + H^+ + 2e^-$$

However, the enzymes involved in this reaction, by which water undergoes light dependent cleavage or "photolysis," have not been isolated.

We can now see the complete pathway taken by electrons arising from the water molecule. They pass from water to chlorophyll *b*, whence, following excitation of the latter, they traverse the connecting electron-transferring chain and enter the empty holes in chlorophyll *a*, from which they are again boosted by illumination and thus pass to $NADP_{ox}$. We can now write the following overall equation

$$2h\nu + H_2O + NADP_{ox} + 2ADP + 2P_i \longrightarrow NADP_{red} + 2ATP + \tfrac{1}{2}O_2 + H_2O$$

The molecule of H_2O that enters is the electron donor for reduction of $NADP_{ox}$ and the source of molecular oxygen; the H_2O molecule appearing on the right-hand side is that extracted from phosphate and ADP to yield ATP. When these are cancelled out we have

$$2h\nu + NADP_{ox} + 2ADP + 2P_i \longrightarrow NADP_{red} + 2ATP + \tfrac{1}{2}O_2$$

6–9 THE MECHANISM OF CYCLIC ELECTRON FLOW AND PHOTOPHOSPHORYLATION

Now we must return to consider the pathway of cyclic electron flow in chloroplasts, under conditions in which $NADP_{ox}$ is absent and no oxygen is being evolved. Can this process also be accounted by the "hookup" in Fig. 6–9, or must some other mechanism be invoked? What is the whole function of cyclic electron flow?

Although the relationship between cyclic and noncyclic electron flows is not yet entirely clear, the hypothesis that is favored by most evidence is that cyclic electron flow occurs in photosystem I by intervention of a "shunt" that makes possible the diversion of electrons from chlorophyll *a* before they reach $NADP_{ox}$, in such a way that they are shunted into the electron chain leading into chlorophyll *a* from photosystem II, as is shown in Fig. 6–9. The shunt must enter at a point so as to include in the closed electron pathway at least one site at which ATP is generated. By means of this shunt, high-energy electrons from excited chlorophyll *a* can return to the latter after a large part of their energy has been tapped off to make ATP, without causing either reduction of $NADP_{ox}$ or evolution of oxygen. But how this shunt is switched in and out is not understood.

An alternative explanation, which has been postulated very recently, is that there is no shunt as just described. Rather it is proposed that chloroplasts of higher plants contain a third photosystem, independent of the linked photosystem I and photosystem II. The sole function of the third photosystem is to generate ATP through cyclic electron flow. The net effect would be that the ratio of ATP formed to $NADP_{ox}$ reduced might vary widely depending on the relative rates at which the third photosystem functions. It thus appears to be the function of light-induced cyclic electron flow to generate ATP at whatever rate is required by the cell, without the necessity of generating simultaneously $NADP_{red}$ or of evolving molecular oxygen.

Whatever the details, the light reactions of photosynthesis cause the formation of three characteristic products: $NADP_{red}$, ATP, and molecular O_2. Now let us see how $NADP_{red}$ and ATP are used in the dark reactions of photosynthesis to bring about the reduction of CO_2 to yield glucose.

6–10 FORMATION OF GLUCOSE IN THE DARK PHASE OF PHOTOSYNTHESIS

The formation of glucose in photosynthesis is a *dark* process that begins with CO_2 and utilizes the $NADP_{red}$ and ATP generated in the light reactions. The overall equation of the dark reactions will be helpful in orienting our discussion. It is

$$6NADP_{red} + 12H_2O + 12ATP + 6CO_2 \longrightarrow C_6H_{12}O_6 + 6NADP_{ox} + 12ADP + 12P_i$$

One of the great biochemical mysteries in photosynthesis was the chemical and enzymatic nature of the first reaction by which the CO_2 molecule is incorporated into a more complex organic form on its way to becoming glucose. This problem was finally elucidated by use of radioactive CO_2 as a tracer in some ingenious experiments carried out by Calvin and his colleagues. Photosynthetic algae were exposed to radioactive CO_2 and then illuminated for very short periods of time. Extracts of the cells were then made and examined to determine which carbon-containing substance first acquired the radioactive carbon atoms of CO_2 on illumination. By such experiments it was found that 3-phosphoglyceric acid, which we have already seen is an intermediate in glycolysis, is among the first compounds to become labeled. Moreover, chemical degradation of the labeled 3-phosphoglyceric acid revealed that the radioactive carbon was largely in the carboxyl group. It was ultimately found that the first step of CO_2 reduction (Fig. 6–10) consists of the

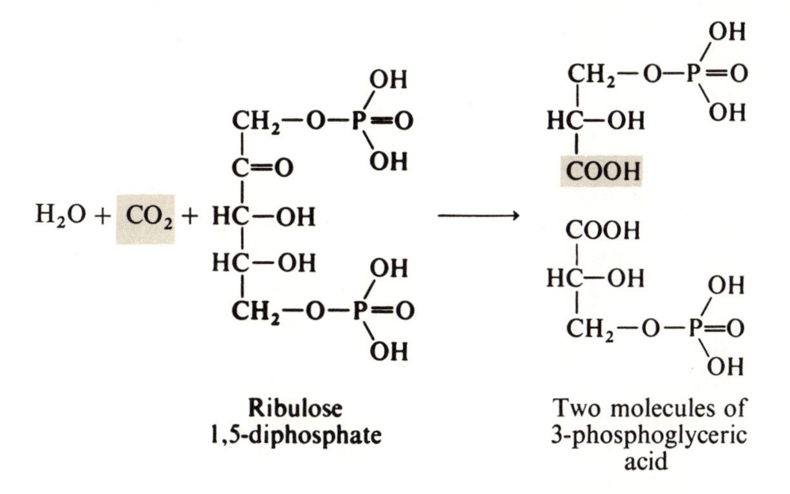

Figure 6–10. The fixation of CO_2 by the ribulose diphosphate carboxydismutase reaction.

reaction of CO_2 with a phosphorylated 5-carbon sugar, *ribulose* 1,5-*diphosphate*, to form two molecules of 3-phosphoglyceric acid. One of these molecules contains the newly incorporated CO_2 (Fig. 6–10).

The two molecules of 3-phosphoglyceric acid so formed now undergo enzymatic reduction to two molecules of 3-phosphoglyceraldehyde, at the expense of $NADP_{red}$ and ATP formed in the light reaction (Fig. 6–11). Then

$$\text{Two 3-phosphoglyceric acid} + 2\text{NADP}_{red} + 2\text{ATP} \longrightarrow \text{two 3-phosphoglyceraldehyde} + 2\text{ADP} + 2\text{P}_i + 2\text{NADP}_{ox}$$

$$\text{Two 3-phosphoglyceraldehyde} \longrightarrow \longrightarrow \text{fructose 1,6-diphosphate}$$

$$\text{Fructose 1,6-diphosphate} + \text{H}_2\text{O} \longrightarrow \text{fructose 6-phosphate} + \text{P}_i$$

$$\text{Fructose 6-phosphate} \longrightarrow \text{glucose 6-phosphate}$$

$$\text{Glucose 6-phosphate} + \text{H}_2\text{O} \longrightarrow \text{glucose} + \text{P}_i$$

Figure 6–11. Formation of glucose from 3-phosphoglyceric acid during dark reactions.

the two molecules of 3-phosphoglyceraldehyde are converted into glucose (Fig. 6–11), essentially by reversal of the reactions of glycolysis (Chapter 4). Note, however, that we have not created the entire glucose molecule from CO_2. We have merely added one carbon atom in the form of a CO_2 molecule to a 5-carbon sugar and ultimately made a 6-carbon sugar of it; this process required the input of two molecules of ATP and two of $NADP_{red}$. This cycle of events is repeated over and over; in each cycle a molecule of CO_2 reacts with the 5-carbon sugar ribulose 1,5-diphosphate to yield two molecules of phosphoglycerate, which then combine to make glucose, at the expense of ATP and $NADP_{red}$ (Fig. 6–12).

Although this scheme may appear to be simple, the pathway of carbon in photosynthesis is actually rather complex. The complexity arises in the enzymatic mechanism by which the 5-carbon sugar ribulose 1,5-diphosphate is regenerated at each revolution of this cycle. In principle, the carbon dioxide reduction cycle is something like the tricarboxylic acid cycle; ribulose 1,5-diphosphate is necessary for each turn and must be regenerated each time, just as oxaloacetate is necessary for citrate formation and must be regenerated in the tricarboxylic acid cycle. The enzymatic reactions required in the regeneration of ribulose 1,5-diphosphate will not be considered in detail.

Recent research strongly suggests that not all species of photosynthetic cells reduce carbon dioxide by the ribulose 1,5-diphosphate pathway. Other pathways have been postulated, particularly for the photosynthetic bacteria, some of which are apparently capable of catalyzing reversal of the tricarboxylic acid cycle reactions, to yield pyruvate from CO_2.

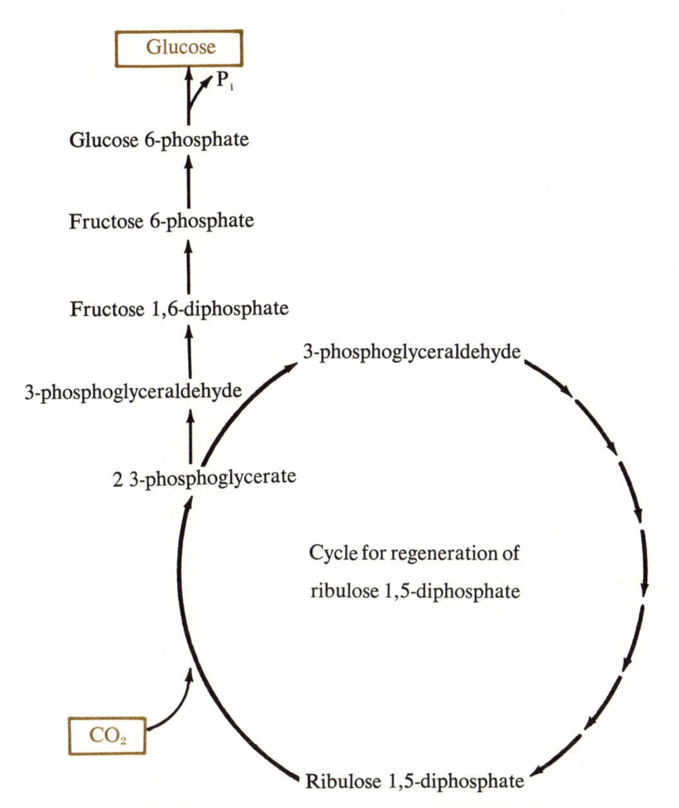

Figure 6–12. Flow plan for dark reactions of photosynthesis. In order to regenerate a molecule of ribulose 1,5-diphosphate in each cycle, a complex sequence of enzyme-catalyzed reactions takes place.

6–11 EFFICIENCY

The question of the thermodynamic efficiency of photosynthesis has long been debated. If we speak of the overall efficiency of photosynthesis in the field, it is not very high, for understandable reasons. For example, it has been estimated from the output of fixed carbon by a field of corn in one growing season, that only 1 or 2 per cent of the energy of the sunlight falling on the field is recovered. However, under laboratory conditions photosynthesizing green cells, such as algae or even isolated chloroplasts, may show a very high efficiency. Many considerations suggest that eight quanta of visible light are required to reduce each molecule of CO_2, or a total of 48 quanta to form one molecule of glucose. One quantum of light in the red end of the visible spectrum ($\sim$ 700 nm) represents about 40 kcal (Fig. 6–2); therefore 48 quanta represents a requirement of 1920 kcal of light energy to make one molecule of glucose. Since the standard free energy of formation of glucose from CO_2 and H_2O is 686 kcal, photosynthesis under laboratory conditions is about $686/1920 \times 100 = 36$ per cent efficient.

6–12 THE CHLOROPLAST

Both the light reactions of photosynthesis, including production of oxygen, ATP, and $NADP_{red}$, as well as the dark reactions leading to glucose formation, take place in the chloroplasts of photosynthetic cells of higher plants. Just as mitochondria are the power plants of heterotrophic cells, chloroplasts are the power plants of photosynthetic autotrophs, at least in the daytime. At night photosynthetic cells become heterotrophic and then utilize their mitochondria for respiration.

Chloroplasts take many forms but often they are ellipsoidal structures from 2 to 8 μ long; they are considerably larger than mitochondria. In cells of higher green plants there may be dozens of chloroplasts in the cytoplasm of each cell, whereas in the smallest photosynthetic eucaryotes, such as some of the higher algae, there may be only one or two. Like the mitochondrion, the chloroplast is surrounded by two membranes (Fig. 6–13). The inner membrane folds inward to form *lamellae*, which are not unlike the cristae of mitochondria. At intervals the lamellae thicken to form structures called *thylakoid disks*. Stacks of these disks are called *grana*. Grana may be isolated after breaking the chloroplast membrane; they contain all the pigments, electron carriers, and phosphorylating enzymes of photosystems I and II. Most of the enzymes required in the dark reactions in which glucose is formed are located in the *stroma*, the semifluid material bathing the lamellae and grana. Figure 6–14 shows a schematic representation of the location of chlorophyll and various important enzymes in the thylakoid disks, which contain many knob-like protrusions when stained under certain conditions. These knobs correspond to the ATP-forming enzymes, which are very similar to those occurring on the inner surface of the inner mitochondrial membrane.

Chloroplasts resemble mitochondria in two other respects. They bring about ion movements when they are illuminated, so that H^+ ions move inward and K^+ and Mg^{2+} move outward. Moreover, the lamellae and grana undergo characteristic changes in shape and structural organization during intermittent light and dark periods. For these reasons, it is believed that the mechanism of photosynthetic phosphorylation in chloroplasts closely resembles the mechanism of oxidative phosphorylation in mitochondria. The physical orientation of the various pigments and electron carriers in the inner membranes is vital to their catalytic function, for the ability of grana to carry out photosynthetic reactions is completely destroyed if membrane structure is disrupted, even by relatively mild procedures.

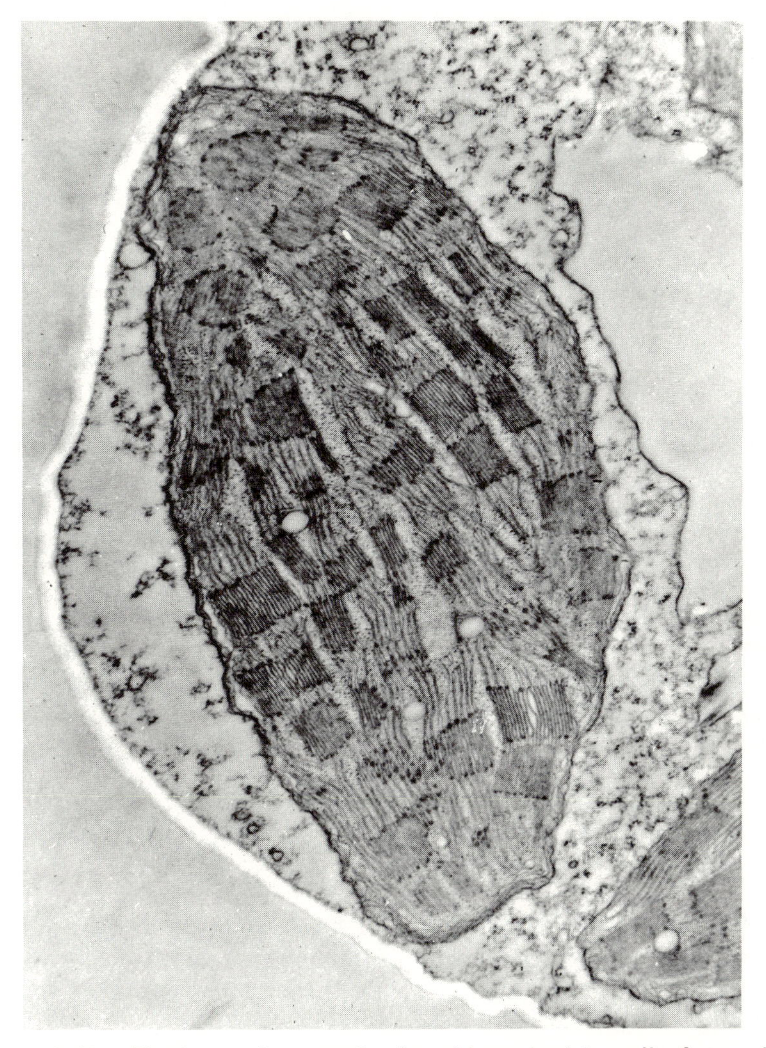

Figure 6–13. Electron micrograph of a chloroplast in cell of corn leaf (courtesy of Dr. Albert E. Vatter, University of Colorado).

We have seen that there are three major hypotheses for the mechanism of oxidative phosphorylation in mitochondria (Section 5–9): the chemical coupling, chemiosmotic coupling, and conformational coupling theories. These have also been postulated for the mechanism of photosynthetic phosphorylation. The chemiosmotic coupling mechanism is favored by most recent research on photophosphorylation. For example, it has been found that when a gradient of hydrogen ions, that is, a pH gradient, is established

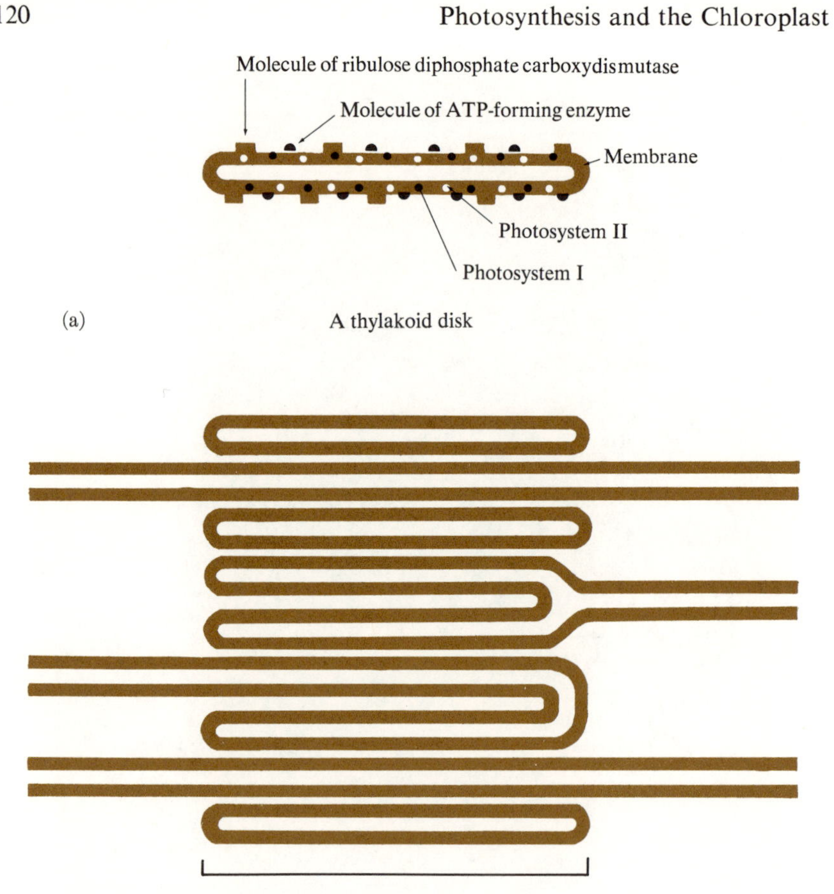

Figure 6–14. Arrangement of the inner membrane of a chloroplast. (a) A representation of the important enzymes and the photosystems in a single thylakoid disk. (b) The manner of folding and stacking of the inner membrane to form grana.

across the thylakoid membrane by artificial means, by making the internal H^+ ion concentration relatively high and the external concentration low, such a gradient can serve as the driving force for the formation of ATP from ADP and phosphate in the dark; as ATP is formed, the gradient of H^+ ions is discharged. It is therefore suggested that photoinduced electron transport serves to generate a gradient of H^+ ions across the membrane.

Clearly the chemistry and physics of the capture of light energy and the relationship of these events to the molecular organization of the chloroplast are among the most challenging and fascinating of all biological problems today.

7

THE CHEMICAL WORK OF BIOSYNTHESIS: POLYSACCHARIDES AND LIPIDS

Biosynthesis is the most complex and vital energy-requiring activity of living organisms. In fact, it is the very essence of the living state, since it includes not only the formation of characteristic chemical components of cells from simple precursors but also their assembly into such structures as membrane systems, contractile elements, mitochondria, nuclei, and ribosomes. Biosynthesis is a genetically programmed process that leads from very simple molecules to the living cell structure itself, in a hierarchy of increasing complexity (Fig. 7–1).

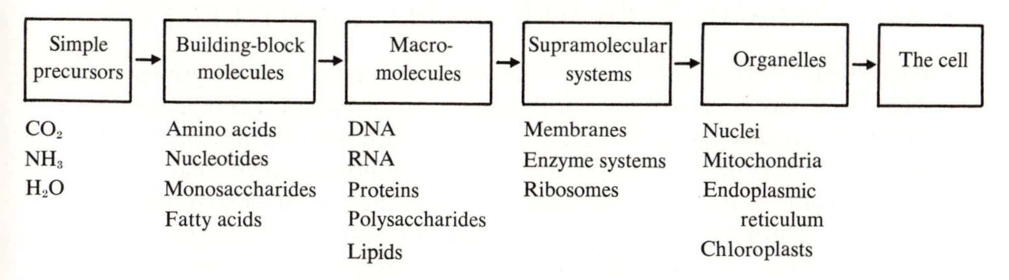

Figure 7–1. Major stages in biosynthesis.

In this chapter we shall examine the dynamics of biosynthesis, its energy requirements, and some representative biosynthetic pathways, namely, those for the synthesis of glucose from simpler precursors and for the assembly of glycogen and phospholipids from their building-block molecules. In succeeding chapters we shall see how the more complex informational macromolecules, namely, proteins and nucleic acids, are assembled from simple precursors, and how macromolecules are in turn assembled into higher order structures.

7–1 CELL GROWTH

Biosynthesis of cell material is most conspicuous in growing organisms because the amount of newly synthesized cell material can be ascertained by the increase in number of cells, as in a culture of multiplying bacteria, or by the increase in weight, as in a growing rat. The rate of growth of cells varies greatly. On one extreme of the biological spectrum, we find that some bacteria, such as *E. coli*, can double in number every 20 minutes. The newly synthesized cellular material is derived entirely from the glucose, ammonium salts, and minerals present in the culture medium. Bacterial cells can, in fact, perform prodigious feats of biosynthesis, as may be seen in Table 7–1. Actually, almost all the metabolic energy of bacterial cells is put into biosynthetic work. Because bacteria normally live in natural environments over which they have little control and from which they cannot escape, their ability to multiply rapidly enables them to survive.

In the more highly organized animal and plant organisms, individual cells generally grow much more slowly. In mammalian organisms we find that neurons of the central nervous system grow only early in the life of the animal, but never divide again during its lifetime. Muscle cells also grow and divide only slowly. Liver cells grow and divide perhaps every two to three months. In contrast, cells of the intestinal mucosa have a relatively rapid rate of growth and division, which may be measured in terms of days or even hours.

7–2 DYNAMIC TURNOVER OF CELL CONSTITUENTS

Biosynthetic work also takes place in cells that are not actively growing in mass. This may seem paradoxical, but many of the chemical components of living cells are undergoing continuous, dynamic turnover. The proteins, lipids, and other components of the cell are built up and broken down continuously in such a way that the rate of synthesis equals the rate of breakdown. The rate of such metabolic turnover of cellular constituents varies considerably from one chemical component of the cell to another. For example, in rat liver cells, which have a life span of some two to three months, the time

TABLE 7–1. Biosynthetic Capacities of a Bacterial Cell
A cell of *Escherichia coli* is about $1 \times 1 \times 3\ \mu$ in size, has a volume of $2.25\ \mu^3$, a total weight of 10×10^{-13} g and a dry weight of 2.5×10^{-13} g. The rates of biosynthesis given were averaged over a 20-minute cell-division cycle. Other simplifying assumptions are described in the text of Chapters 10 and 11.

Chemical component	Per cent of dry weight	Approximate molecular weight	Number of molecules per cell	Number of molecules synthesized per second	Number of molecules of ATP required to synthesize per second	Per cent of total biosynthetic energy required
DNA	5	2,000,000,000	1	0.00083	60,000	2.5
RNA	10	1,000,000	15,000	12.5	75,000	3.1
Protein	70	60,000	1,700,000	1400	2,120,000	88.0
Lipids	10	1,000	15,000,000	12,500	87,500	3.7
Polysaccharides	5	200,000	39,000	32.5	65,000	2.7

required for half of the glycogen of the cell to be replaced (i.e., the *half-time*) is less than six hours, the half-time of the phospholipids about three days, and that of the proteins about seven to ten days. Even the internal organelles of the cell undergo metabolic turnover within the life span of the parent cell. Recent research has shown that the mitochondria of the liver cell have a half-time of only five to ten days. Moreover, in rapidly growing cells of the mold *Neurospora crassa*, the mitochondria are built from very simple precursor molecules in much less than two hours.

The rate of metabolic turnover of cell components also varies considerably from one cell type to another. Whereas the proteins of the liver cell have a half-period of about ten days, those of skeletal muscle have a half-time of some three months. In general, the more complex and differentiated the cell, the slower is its overall rate of turnover, although some individual components of such cells may turn over rapidly.

Chemical work of biosynthesis is also involved in the output of certain characteristic products that are secreted by the cell and deployed in its environment. For example, the *fibroblast* is a specialized connective tissue cell, in higher animals which makes and extrudes vast amounts of *collagen*, a fibrous, insoluble, structural protein. Actually, collagen is the most abundant protein in the mammalian body; its fibers serve as an extracellular structural material to strengthen and support all tissues and organisms. The *acinar cells* of the pancreas also manufacture large amounts of protein for export. They secrete into the small intestine the enzymes trypsin, chymotrypsin, and carboxypeptidase, as well as lipase and ribonuclease. These enzymes catalyze the breakdown of proteins, lipids, and nucleic acids of the ingested food in preparation for its absorption into the blood stream.

7–3 RATE OF BIOSYNTHESIS AND THE DISTRIBUTION OF BIOSYNTHETIC ENERGY

Now let us consider in more quantitative terms the amount of work required to construct each of the major molecular components of the cell from its basic precursor molecules, in terms of the ATP input, to get some idea of the rate and scope of biosynthetic reactions and of the fraction of the total biosynthetic ATP energy that is required for construction of each of the major types of cell components.

In Table 7–1 are given some approximate figures on the major chemical components of a single *E. coli* cell. Also given are the approximate molecular weights of these components, which must be regarded as average values, simplified for purposes of calculation. For example, the protein molecules of *E. coli* cells exist in a spectrum of molecular weights, which may range from as little as 13,000 to as high as 1,000,000; the assumed average molecular weight of 60,000 is a reasonable approximation. Similarly, there are at least three major kinds of RNA molecules in *E. coli* cells that differ widely in molecular weight, but again, it is not unreasonable to use the average value of 1,000,000. The number of molecules of each type in the cell is easily calculated from these data and from Avogadro's number (6.02×10^{23}), which is the number of molecules in a gram molecular weight of any compound. The average number of molecules of each component synthesized per second can then be calculated, if it is assumed that each cell requires 20 minutes for its complete synthesis and that the biosynthesis proceeds linearly with time over the 20-minute interval. From information we shall develop later in this chapter and in the next, it is also possible to calculate how many molecules of ATP are required to build each molecular component of the cell in each 20-minute growth cycle. We see, for example, that eight molecules of ATP are required to build a typical phospholipid molecule, phosphatidyl choline, from its five building blocks, namely, glycerol, phosphoric acid, choline, and two molecules of fatty acids. Finally, it is of considerable interest to calculate how the ATP energy is distributed in the biosynthesis of the various cell components.

Let us now examine some of the data in Table 7–1. First we note that proteins are the most abundant organic substances in the cell; they make up well over 50 per cent of the total dry weight; nucleic acids, lipids, and polysaccharides follow in that order. Next we observe the prodigious rate of biosynthesis of the different cell components. The most impressive biosynthetic performance is, however, the synthesis of protein molecules, of which some 1400 may be completed per second. Although more phospholipid molecules are synthesized per second than protein molecules, phospholipids are small molecules whose synthesis requires the formation of only four new covalent bonds, or about 5600 new covalent bonds per second. On the other hand each protein molecule contains over 300 new covalent linkages, on the average, so that protein synthesis by *E. coli* requires formation of $300 \times 1400 = 420{,}000$ new covalent linkages per second. Moreover, the twenty different kinds of amino acids found in proteins are arranged in specific sequences during their biosynthesis by the genetic coding system, which we shall describe in Chapter 8. Let us now compare this performance with the efforts of organic chemists in the laboratory. Only recently has a protein molecule been synthesized in the laboratory for the first time by nonbiological procedures. This protein, ribonuclease, has a chain of 129 amino acid units in an ordered sequence. However, this very creditable feat required months of preparation, many highly skilled chemists, and much elaborate equipment.

The data in Table 7–1 also permit another important conclusion. The synthesis of proteins is quantitatively by far the most important biosynthetic activity of the cell. Protein synthesis requires up to 90 per cent of the available ATP used for biosynthesis in the *E. coli* cell; the formation of lipids, nucleic acids, and polysaccharides requires but relatively minor fractions of the total biosynthetic energy. Moreover, it is also clear that the biosynthesis of all the different cell components must be rigidly controlled and integrated, so as to yield the various proteins, lipids, nucleic acids, and polysaccharides in the proper molar ratio characteristic of an *E. coli* cell.

Actually, the figures recorded in Table 7–1 are but *minimum* statements of the total ATP energy required for biosynthetic activities of the cell. In the first place, these calculations assume that in each *E. coli* cell all the protein, nucleic acid, lipid, and polysaccharide molecules are built only once in the 20-minute lifetime of the cell. Second, it must be pointed out that the calculations of ATP requirements given are based on the assumption that the cell is already supplied with preformed building blocks, that is, amino acids for protein synthesis, mononucleotides for nucleic acid synthesis, sugar molecules for polysaccharide synthesis, and fatty acids and other units for lipid synthesis. But *E. coli* cells can also build their own building blocks from even simpler precursors. For example, they can build all the amino acids required for protein synthesis and all the nucleotides for nucleic acid synthesis from simple compounds like ammonium salts and glucose. Construction of these building-block molecules also requires input of large amounts of ATP.

To sum up, the biosynthetic work carried out by the cell is probably its most complex and profound energy-requiring activity. Biosynthetic work encompasses the most fundamental biological phenomenon known, and includes the preservation and transmission of genetic information, the transcription of this information into specific cellular components having characteristic functional activities, and finally the differentiation of cell structure of different organisms.

Now that we have seen the scope and rate of biosynthetic activities in cells, let us return to some first principles as we approach the biochemical organization of biosynthetic processes.

7–4 THE FLOWSHEET OF BIOSYNTHESIS

For the biosynthesis of the various cell components two kinds of ingredients are required: (1) precursors that provide the carbon, hydrogen, nitrogen, and other atoms characteristic of the final biosynthetic product, and (2) ATP and other forms of chemical energy required to assemble the precursors into the covalently-bonded structure of the product.

First let us examine the chemical nature of the precursors for biosynthesis and survey the major pathways by which biosynthesis occurs. As examples let us take heterotrophic cells, of which *E. coli* or the liver cells of mammals are good examples. These, as well as all other heterotrophic cells, must obtain the carbon required for biosynthesis in a form other than carbon dioxide; usually they obtain it as glucose, although they may use other nutrients, such as fatty acids, as a carbon source. They also require a source of nitrogen. The needs of

E. coli cells for nitrogen are satisfied by ammonia (NH_3), from which they can make all the amino acids required for synthesis of protein. However, liver cells of mammals can make only about half of the required amino acids from ammonia; the rest must be provided preformed in the dietary intake of the animal. Most heterotrophic cells can in fact manufacture all their cell components starting from but two bulk nutrients: glucose and either ammonia or amino acids.

How can cells manufacture such a diversity of different cell components from only two major precursors? The answer lies in the design of the central metabolic pathways of cells, which interconnect in such a way that carbon atoms originally derived from glucose can be used to build not only glycogen but also fatty acids, the carbon skeleton of amino acids, and the carbon portions of nucleotides. Figure 7–2 shows how the various metabolic pathways of the cell are arranged. The hundreds of enzymes in the cell do not operate independently but rather function in many multienzyme sequences, which are interconnected so that during catabolism or degradation of cell components, all the products ultimately flow toward a central common pathway and during biosynthesis a few precursors are built up into many products by diverging or branching central pathways.

We see from Fig. 7–2 that the breakdown and biosynthesis of cell components proceeds in three major stages. Moreover, we see that stage 3 of metabolism represents a common central or main-trunk pathway common to both breakdown and biosynthetic reactions of metabolism. Thus, from only a relatively few precursor molecules, derived from stage 3 a vast variety of different cell components can be synthesized, including the hundreds of different proteins of the cell, the numerous types of lipids, many nucleic acids, and several types of polysaccharides.

7–5 THE BIOSYNTHESIS OF GLUCOSE FROM SIMPLE PRECURSORS

Another important feature illustrated by the blueprint or flowsheet of metabolism in Fig. 7–2 is that the degradative pathways and the biosynthetic pathways between two given points, such as between glucose and pyruvic acid, are not identical, as is indicated by the two different colors used for the arrows. This duality of the central metabolic pathways is a thermodynamic necessity, since we know degradative reactions proceed with a decrease in free energy whereas biosynthetic reactions require the input of free energy.

Let us take one of the most important central routes of metabolism, namely, the path between glucose to pyruvic, to see how the degradative and synthetic pathways between these two metabolites differ in pattern and energetics. The simplest way is to show these opposite pathways schematically. We see in Fig. 7–3 that the degradation of glucose to pyruvic acid proceeds through the

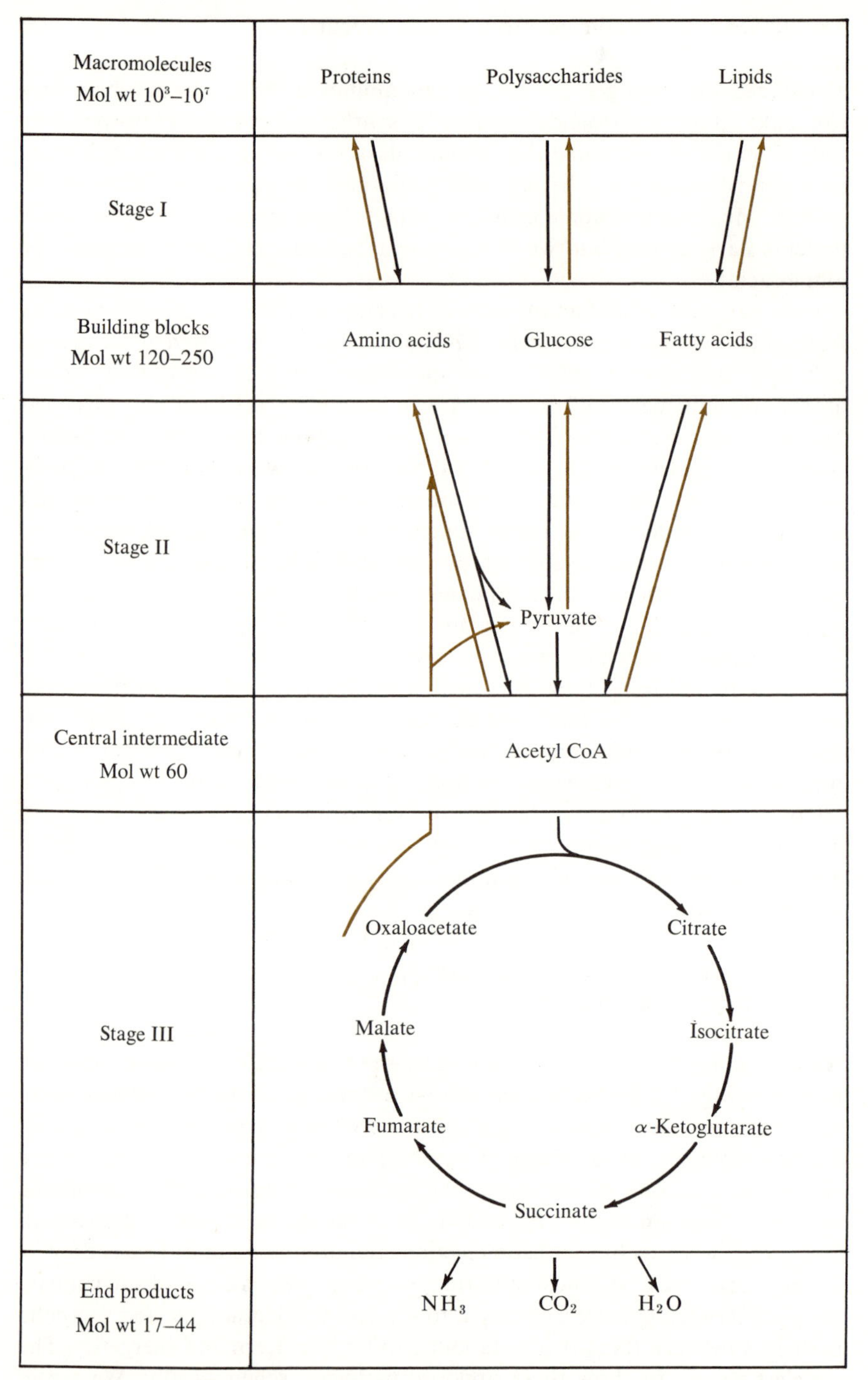

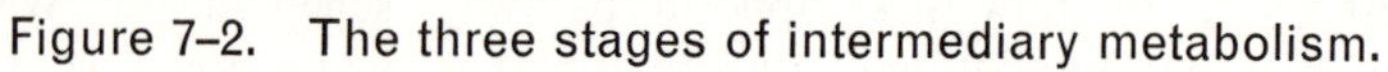

Figure 7–2. The three stages of intermediary metabolism.

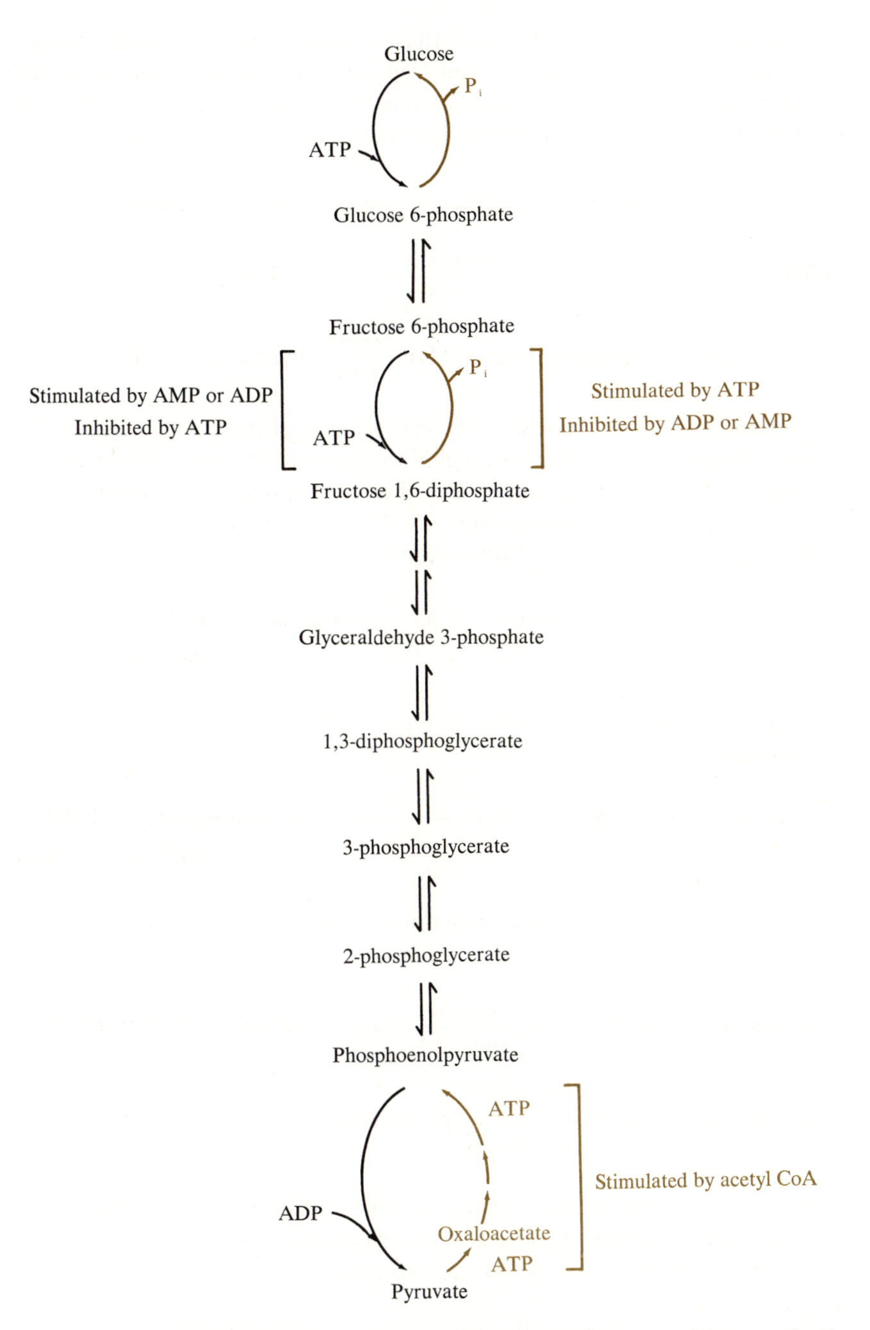

Figure 7–3. Independent regulation of the degradative and biosynthetic pathways between glucose and pyruvate.

various successive steps of glycolysis that we have already examined (Section 4–4). Although seven of the reaction steps of glycolysis are freely reversible and are used both for the downhill and uphill pathways, there are three reactions that are essentially irreversible in the direction of degradation

$$\text{ATP} + \text{glucose} \xrightarrow{\text{hexokinase}} \text{glucose 6-phosphate} + \text{ADP}$$

$$\text{ATP} + \text{fructose 6-phosphate} \xrightarrow[\text{kinase}]{\text{phosphofructo-}} \text{fructose 1,6-diphosphate} + \text{ADP}$$

$$\text{phosphoenolpyruvate} + \text{ADP} \xrightarrow[\text{kinase}]{\text{pyruvate}} \text{pyruvate} + \text{ATP}$$

These reactions cannot reverse in the cell because of their large negative free energy changes, that is, they are highly exergonic. However, these reactions are bypassed during biosynthesis by other enzyme-catalyzed reactions *that are exergonic in the direction of synthesis.* The hexokinase reaction is replaced by that catalyzed by *glucose 6-phosphatase*

$$\text{glucose 6-phosphate} + H_2O \longrightarrow \text{glucose} + H_3PO_4 \qquad \Delta G^{0\prime} = -3.3 \text{ kcal}$$

the phosphofructokinase reaction is replaced by that catalyzed by *fructose diphosphatase*

$$\text{fructose 1,6-diphosphate} + H_2O \longrightarrow \text{fructose 6-phosphate} + H_3PO_4 \qquad \Delta G^{0\prime} = -4.0 \text{ kcal}$$

and the pyruvate phosphokinase reaction is replaced by a series, of which the two key reactions are

$$\text{pyruvic acid} + \text{ATP} + CO_2 \xrightarrow[\text{carboxylase}]{\text{pyruvate}} \text{oxaloacetic acid} + \text{ADP} + H_3PO_4$$

$$\text{oxaloacetic acid} + \text{GTP} \underset{\text{carboxykinase}}{\overset{\text{phosphoenol pyruvate}}{\rightleftharpoons}} \text{phosphoenolpyruvic acid} + CO_2 + \text{GDP}$$

Now let us write out the total or overall equations for the degradative pathway from glucose to pyruvic acid and its counterpart, the equation for the biosynthetic conversion of pyruvic acid to glucose

Degradative pathway

$$\text{glucose} + 2\text{ADP} + 2P_i + 2\text{NAD}_{ox} \longrightarrow 2 \text{ pyruvic acid} + 2\text{ATP} + 2\text{NAD}_{red} + 2H_2O$$

Synthetic pathway

$$2 \text{ pyruvic acid} + 4ATP + 2GTP + 2NAD_{red} + 6H_2O \longrightarrow \text{glucose} + 2NAD_{ox} + 4ADP + 2GDP + 6P_i$$

We can see not only that the degradative and biosynthetic pathways are not the simple reverse of each other but also that they differ with respect to their energetics. Degradation of one molecule of glucose to pyruvate is accompanied by formation of two high-energy phosphate bonds of ATP, whereas the synthesis of a molecule of glucose from pyruvate requires input of a total of six high-energy phosphate bonds, four from ATP and two from GTP. Both overall reactions are essentially irreversible in the direction they are written.

There is one additional important feature of the dual pathways of degradation and biosynthesis: they are independently regulated in the cell. We have already seen that the major control point in glycolysis, that is, the pacemaker reaction, is the phosphorylation of fructose 6-diphosphate to yield fructose 1,6-diphosphate. We now know that there are two major control points in the biosynthetic pathway. One is the reaction by which fructose 1,6-diphosphate is hydrolyzed to fructose 6-phosphate, which is catalyzed by a regulatory enzyme whose positive modulator is ATP and whose negative modulator is AMP or ADP. The other control point is the conversion of pyruvic acid to oxaloacetic acid, which is activated by acetyl CoA. Thus, whenever the ATP and acetyl CoA concentrations in the cell exceed certain levels, the degradative pathway is inhibited and the biosynthetic pathway is stimulated.

7–6 ENERGY RELATIONSHIPS IN THE ASSEMBLY OF BUILDING BLOCKS INTO MACROMOLECULES

Now let us see how cells assemble the different characteristic building blocks into macromolecules. The formation of large, complex macromolecules of the cell by assembly of many building-block molecules is a process in which entropy decreases. For example, a large molecule of glycogen, containing 500 identical glucose units in a covalently bonded chain, is a more organized, less random system than 500 free glucose molecules in a dilute aqueous solution, in which they are randomly disposed and free to change their positions. The synthesis of large molecules of proteins from amino acid building blocks, of which there are twenty kinds, proceeds with an even greater increase of order and decrease of entropy than that of glycogen, because the amino acid units are built into a specific sequence by the cell. From these examples it is clear that biosynthetic reactions of the cell are not spontaneous reactions, since they proceed with a decrease in entropy and thus an increase in free energy. They can therefore occur only if they are coupled to other reactions that can furnish the necessary free energy.

The second matter that deserves special mention is that the assembly of all the various macromolecules from simple building blocks involves the removal of the elements of water from the successive units, to form the glycosidic bonds of polysaccharides, the peptide bonds of proteins, and the phosphodiester bonds of the phospholipids and nucleic acids. These reactions take place in the dilute aqueous medium of the cell. We have seen that the hydrolysis of such bonds proceeds with a large decrease in free energy; accordingly their synthesis in dilute aqueous systems requires a large input of free energy.

We have already seen how the principle of the common intermediate makes possible the conservation of oxidation-reduction energy as the phosphate bond energy of ATP in glycolysis and respiration. This principle is also involved in the biosynthetic reactions of the cell; in fact it is in the performance of chemical work at the expense of ATP energy that we can most clearly see the principle of the common intermediate in operation. In Chapter 3 we examined the mechanism of a very simple ATP-dependent biosynthetic reaction, namely, the formation of glutamine from glutamic acid and ammonia (Section 3–10). By the enzymatic transfer of a phosphate group from ATP to glutamic acid, glutamyl phosphate is formed. In the second step, glutamyl phosphate reacts enzymatically with ammonia to produce glutamine, and this equilibrium is such that glutamine can readily accumulate in a dilute aqueous system

$$\text{ATP} + \text{glutamic acid} \rightleftharpoons \text{ADP} + \boxed{\text{glutamyl phosphate}}$$

$$\boxed{\text{glutamyl phosphate}} + \text{NH}_3 \longrightarrow \text{glutamine} + \text{phosphate}$$

$$\text{Sum: ATP} + \text{glutamic acid} + \text{NH}_3 \longrightarrow \text{glutamine} + \text{ADP} + \text{phosphate}$$

$$\Delta G^{0\prime} = -3.90 \text{ kcal}$$

Thus the chemical energy of ATP is used to cause the coupled synthesis of glutamine through the common intermediate glutamyl phosphate, which is shaded in the equation shown above. In all biosynthetic reactions of this type, in which formation of a new linkage is coupled to breakdown of ATP, the overall $\Delta G^{0\prime}$ is negative; this is a thermodynamic necessity to insure that the biosynthetic reaction proceeds essentially to completion.

The common intermediate principle is used in all biosynthetic reactions in the cell in which building block molecules are linked by condensation reactions, that is, those in which H_2O is removed to create the linkage. However, the enzymatic pathways by which polysaccharides, lipids, and proteins are built are usually more elaborate than in the simple case of glutamine, because of an important feature of some ATP-linked reactions that we have not yet encountered. In all our preceding discussions of the principle of the common intermediate in energy transfer reactions involving ATP, we have stressed the importance of the transfer of the terminal phosphate group of ATP to the building-block molecule to be "energized," and the rephosphorylation of the ADP so formed at the expense of the phosphate group of a high-energy phosphate donor. But many biosynthetic reactions for which ATP provides the driving force do not proceed by loss of the single terminal phosphate group of ATP, but rather by loss of the two terminal phosphate groups in one piece as *pyrophosphate*; AMP rather than ADP is the product of such reactions. These two types of cleavage of ATP are contrasted in the following equations:

Orthophosphate cleavage

$$\underset{\text{ATP}}{Ⓐ—Ⓡ—P\sim P\sim P} + H_2O \longrightarrow \underset{\text{ADP}}{Ⓐ—Ⓡ—P\sim P} + \underset{\text{orthophosphate}}{P}$$

Pyrophosphate cleavage

$$\underset{\text{ATP}}{Ⓐ—Ⓡ—P\sim P\sim P} + H_2O \longrightarrow \underset{\text{AMP}}{Ⓐ—Ⓡ—P} + \underset{\text{pyrophosphate}}{P\sim P}$$

The standard free energy change is about the same for the two types of ATP hydrolysis. Pyrophosphate, however, cannot be used directly by the cell to participate in phosphorylation of ADP during glycolysis or oxidative phosphorylations. It must first be enzymatically hydrolyzed to orthophosphate by the enzyme *pyrophosphatase*

$$^{-}O-\overset{\overset{\large O^-}{|}}{\underset{\underset{\large O}{\|}}{P}}-O-\overset{\overset{\large O^-}{|}}{\underset{\underset{\large O}{\|}}{P}}-O^- + H_2O \longrightarrow HO-\overset{\overset{\large O^-}{|}}{\underset{\underset{\large O}{\|}}{P}}-O^- + HO-\overset{\overset{\large O^-}{|}}{\underset{\underset{\large O}{\|}}{P}}-O^-$$

$$\Delta G^{0\prime} = -7.3 \text{ kcal}$$

The hydrolysis of pyrophosphate proceeds with a large negative standard free energy change; pyrophosphate is thus a "high-energy" phosphate compound. Those biosynthetic reactions that result in cleavage of pyrophosphate from ATP thus ultimately lead to the hydrolysis of *two* high-energy phosphate bonds of ATP. This process evidently results in a much larger thermodynamic "pull" to such coupled reactions (i.e., 2×7.3 or 14.6 kcal/mole) than is the case for reactions in which only the single terminal phosphate group of ATP is lost. The reactions involving pyrophosphate cleavage also employ the common intermediate principle, but in a somewhat more complex way than occurs in the particular sequence of reactions shown above for synthesis of glutamine. We shall presently see that pyrophosphate is the cleavage product of nucleoside triphosphates in the case of the synthesis of glycogen and lipids, of DNA and RNA, and of proteins. Pyrophosphate formation appears in all

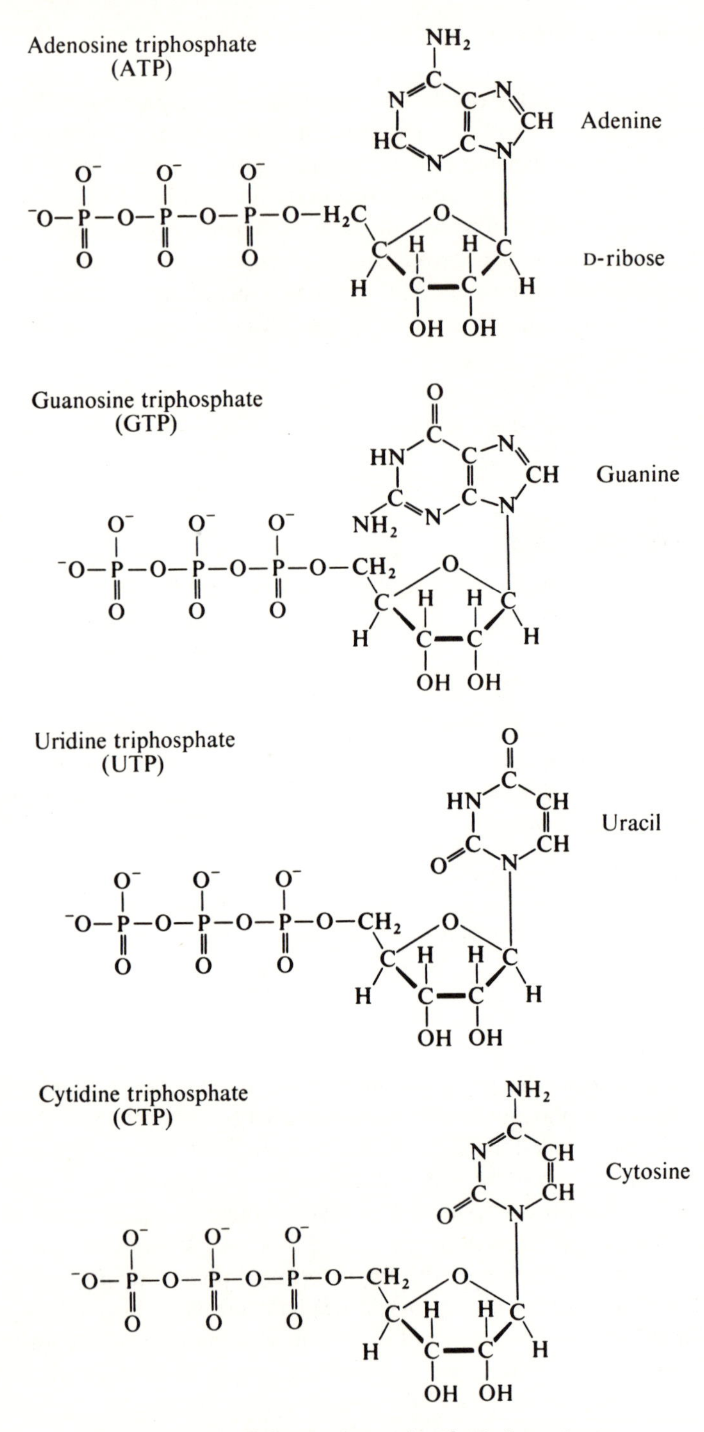

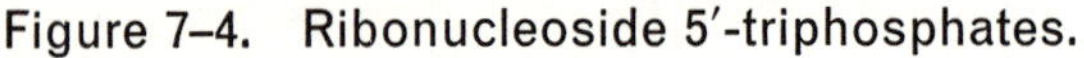
Figure 7–4. Ribonucleoside 5′-triphosphates.

cases to be a chemical device to ensure that these biosynthetic reactions proceed essentially to completion.

7–7 CHANNELING OF ATP ENERGY BY WAY OF OTHER TRIPHOSPHATES

The compounds ATP, ADP, and AMP are of universal occurrence in all cells; they are obligatory components in phosphate group transfer from energy-yielding respiration to the various energy-requiring processes in cells. However, cells contain other phosphate compounds very similar in structure to ATP, which participate as switching or channeling elements in the transfer of phosphate bond energy. These compounds occur in two series. The four *ribonucleoside 5'-triphosphates* constitute the first group (Fig. 7–4). These are analogous with ATP; the adenine ring of ATP is replaced by the purine *guanine* or by the pyrimidines *uracil* or *cytosine*. They are named and abbreviated in a very similar way: *guanosine triphosphate* is GTP, *guanosine diphosphate* is GDP, and so on. The second group consists of four *2-deoxyribonucleoside 5'-triphosphates*, namely, *deoxyadenosine triphosphate* (dATP), *deoxyguanosine triphosphate* (dGTP), *deoxythymidine triphosphate* (dTTP), and *deoxycytidine triphosphate* (dCTP). These compounds are analogous to the ribonucleoside triphosphates, but they contain instead of ribose its derivative 2-deoxyribose

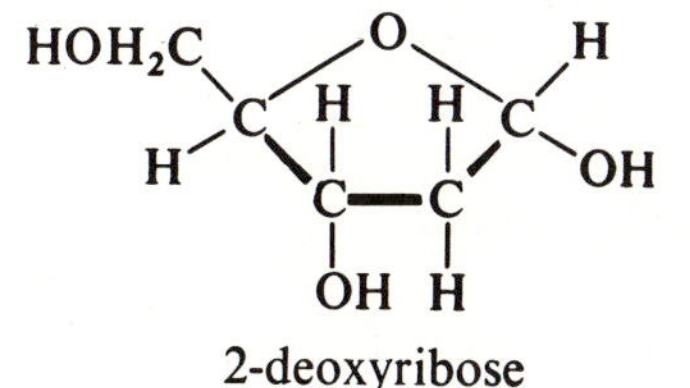

2-deoxyribose

In the deoxyribose series uracil is replaced by the pyrimidine thymine; dTTP has the structure

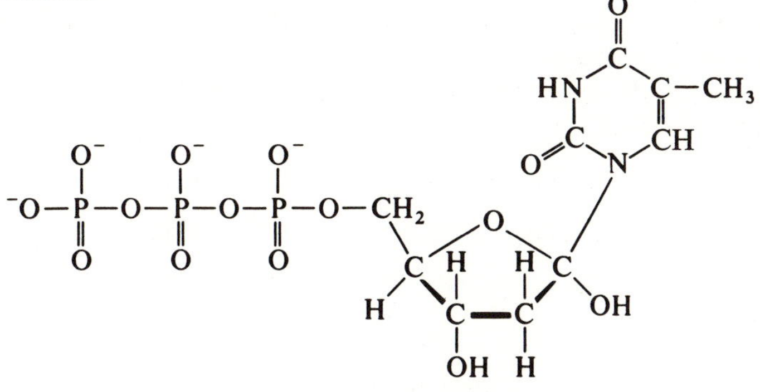

Deoxythymidine triphosphate

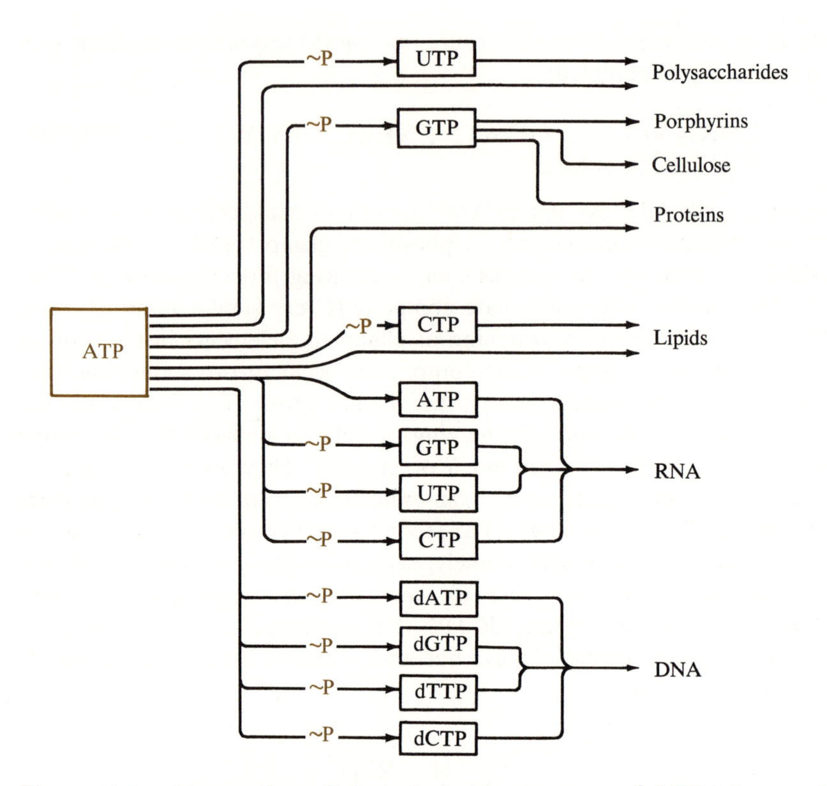

Figure 7–5. Channeling of phosphate bond energy of ATP into specific biosynthetic routes.

Adenosine triphosphate (ATP) is the obligatory intermediate in transfer of bond energy, whereas all the other nucleoside 5′-triphosphates and deoxynucleoside 5′-triphosphates function to channel ATP energy into different biosynthetic pathways (Fig. 7–5). Such a channeling process is made possible because the terminal phosphate group of ATP can be enzymatically transferred to the 5′-diphosphates of the other nucleosides by the action of enzymes called *nucleoside diphosphokinases.* They catalyze the following types of reversible reactions:

$$\text{ATP} + \text{GDP} \rightleftharpoons \text{ADP} + \text{GTP}$$

$$\text{ATP} + \text{UDP} \rightleftharpoons \text{ADP} + \text{UTP}$$

$$\text{ATP} + \text{CDP} \rightleftharpoons \text{ADP} + \text{CTP}$$

$$\text{ATP} + \text{dADP} \rightleftharpoons \text{ADP} + \text{dATP}$$

$$\text{ATP} + \text{dGDP} \rightleftharpoons \text{ADP} + \text{dGTP}$$

and so on.

The compounds GTP, UTP, CTP, and dATP, as well as all the other ribo- or deoxyribo-analogs of ATP, have the same standard free energy of hydrolysis of the terminal phosphate group as ATP; therefore the equilibrium constant of these reactions is approximately 1.0. Because only ADP can accept phosphate groups during oxidative phosphorylation or glycolysis, the ATP-ADP system is necessary to phosphorylate GDP, UDP, CDP, and so on, and thus to fill each channel with high-energy phosphate groups. Each chemical channel then funnels high-energy phosphate groups into the biosynthesis of different types of macromolecules (Fig. 7–5). Uridine 5′-triphosphate (UTP) is the immediate energy carrier in biosynthesis of most polysaccharides in animal tissues; cytidine 5′-triphosphate (CTP) is a specific energy carrier in reactions of lipid biosynthesis, and GTP functions as energy donor in biosynthesis of proteins and other compounds. However, ATP ultimately provides the phosphate groups for all biosynthetic reactions, as well as for active transport and contractile activity (Fig. 7–5). It appears most probable that ATP occupies the central role in energy-transfer reactions because it may have been the first nucleoside 5′-triphosphate to have been formed in the prebiotic era and thereby was selected for the role of energy carrier in the first primitive cells.

7–8 BIOSYNTHESIS OF GLYCOGEN FROM GLUCOSE

Let us now examine the enzymatic pathways and the energetics of synthesis of the major chemical components of the cell. Of these, the polysaccharides are best to begin with. They consist of long chains of simple sugar molecules joined in glycosidic linkage. They may have molecular weights as high as ten million. Among the polysaccharides of wide occurrence are *amylose*, the major component of starch, which is a storage form of glucose in plants. In animal cells the counterpart storage polysaccharide is *glycogen*. Amylose and glycogen consist of chains of glucose molecules joined in 1,4-glycosidic linkage; they contain no other types of building blocks. Amylose has long, unbranched chains arranged in a helical, coiled structure, whereas glycogen has a highly branched structure (Fig. 7–6). Other polysaccharides include *cellulose*, an insoluble fibrous compound that serves in plants the same structural function served by collagen in higher animals, and *hyaluronic acid*, a representative of the *acid mucopolysaccharides*, a group of viscous, sticky macromolecules that are often found as a lubricant or coating on cells and as components of the extracellular connective tissue between or among cells.

Now let us examine the enzymatic biosynthesis of glycogen in detail because it illustrates the pattern involved in the biosynthesis of most polysaccharides. Figure 7–6 shows schematic representations of the structure of glucose, the building block of glycogen, of the 1,4-glycosidic linkages between successive glucose units, and of the structure of glycogen. Glycogen is enzymatically synthesized from glucose by a repetitive process in which single glucose molecules are attached to the growing end of the glycogen chain. To attach each glucose unit, a series of six reactions is required, each catalyzed by a specific enzyme. The full sequence is given by the reactions in Table 7–2. First glucose is phosphorylated by ATP to yield glucose 6-phosphate, thus activating or raising its energy content. The glucose 1-phosphate formed in the next step now reacts with uridine triphosphate (UTP) to form the immediately reactive carrier of the glucose molecule, namely, *uridine diphosphate glucose*, abbreviated UDP-glucose (Fig. 7–6). This compound may be looked upon as a carrier for glucose, just as ATP is a carrier of phosphate groups. In the next step, catalyzed by the enzyme *glycogen synthetase*, the UDP-glucose reacts with a free end of an existing glycogen chain having *n* glucose units. The glucose molecule is transferred to the end of the chain, lengthening it by one unit, and UDP is left behind. UTP is then regenerated from UDP by enzymatic transfer of the terminal phosphate group from ATP. This entire cycle of reactions must be repeated for each glucose unit added to the chain.

Two important points may now be noted. The first is that glucose 1-phosphate must first be converted by UTP to a derivative of UDP just prior to the

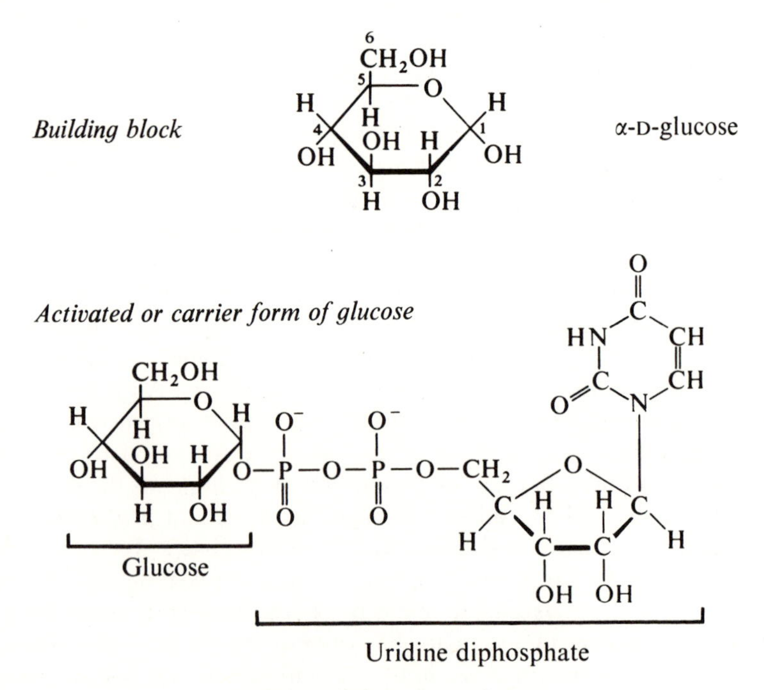

Figure 7–6. Elements in the biosynthesis of glycogen.

Figure 7-6—*continued*

Major chemical linkage between glucose units

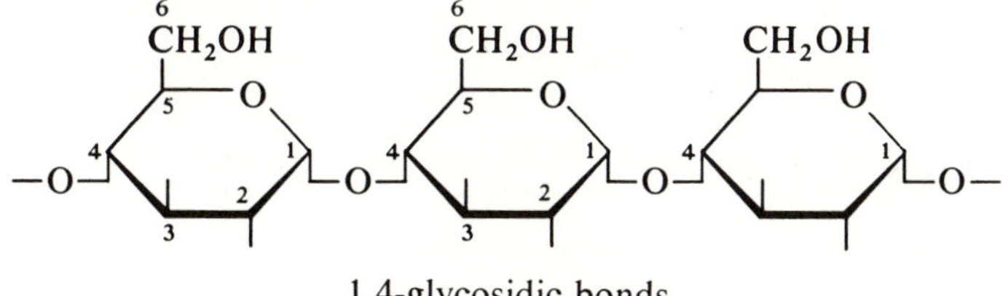

1,4-glycosidic bonds

Glycogen

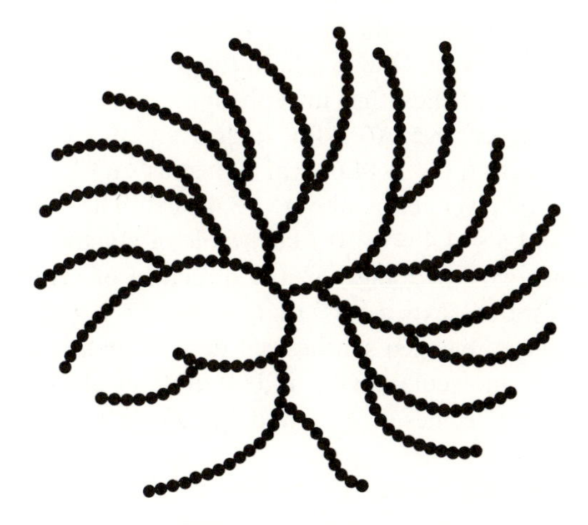

Each black circle represents a glucose residue. The molecular weight of glycogen may exceed one million.

reaction forming the glycosidic linkage. No other nucleoside triphosphate will replace UTP in this reaction. Furthermore, the biosynthesis of most other important polysaccharides and sugar derivatives in animal tissues, such as hyaluronic acid, also involves UDP as the carrier of the monosaccharide building block. In plants, ADD or GDP may serve as the specific carrier in synthesis of cellulose.

TABLE 7–2. The Enzymatic Steps in the Biosynthesis of Glycogen

1. Glucose + ATP $\xrightarrow{\text{hexokinase}}$ glucose 6-phosphate + ADP

2. Glucose 6-phosphate $\xrightarrow{\text{phosphoglucomutase}}$ glucose 1-phosphate

3. Glucose 1-phosphate + UTP $\xrightarrow[\text{pyrophosphorylase}]{\text{UDP-glucose}}$ UDP-glucose + pyrophosphate

4. UDP-glucose + glycogen_n $\xrightarrow[\text{synthetase}]{\text{glycogen}}$ glycogen_{n+1} + UDP

5. ATP + UDP $\xrightarrow[\text{diphosphokinase}]{\text{nucleoside}}$ ADP + UTP

6. Pyrophosphate + H_2O $\xrightarrow{\text{pyrophosphatase}}$ 2 phosphate

Sum: glucose + 2ATP + glycogen_n $\longrightarrow$ 2ADP + 2P + glycogen_{n+1}

The second point is that altogether *two* molecules of ATP are split to ADP and phosphate for each glycosidic linkage formed, as can be seen in the overall equation of the full cycle. One is split in the phosphorylation of glucose to glucose 6-phosphate and the other in rephosphorylating UDP to UTP. The standard free energy change $\Delta G^{0\prime}$ for hydrolysis of two moles of ATP to ADP and P is $2 \times -7.3 = -14.6$ kcal, whereas $\Delta G^{0\prime}$ for synthesis of a glycosidic linkage of glycogen is about +3.4 kcal. The net thermodynamic driving force is $-14.6 + 4.4 = -10.2$ kcal/mole of glucose. This force causes the reaction to go overwhelmingly in the direction of synthesis. This large input may appear more "expensive" than necessary, but this is the price the cell must pay to have glycogen synthesized in a dilute aqueous system. Synthesis of a glycogen molecule having 10,000 units of glucose would require the input of 20,000 molecules of ATP and 10,000 cycles of the six enzymatic reaction steps shown.

Other polysaccharides, such as amylose, cellulose, and hyaluronic acid, are assembled from simple sugars in very similar reactions requiring sugar nucleotides as intermediates.

7–9 BIOSYNTHESIS OF A LIPID FROM ITS BUILDING BLOCKS

Now let us examine the biosynthesis of a typical lipid. The lipids as a class are actually much smaller molecules than polysaccharides; their molecular weights are rarely over 1000. However, their biosynthesis is more complex than that of glycogen because most of them contain more than one type of chemical linkage, whereas glycogen has only one basic type of chemical linkage uniting the glucose units.

The simplest lipids are the *neutral lipids* or *triacylglycerols* (also called triglycerides); these are the storage fats found in adipose or fatty tissue. Their

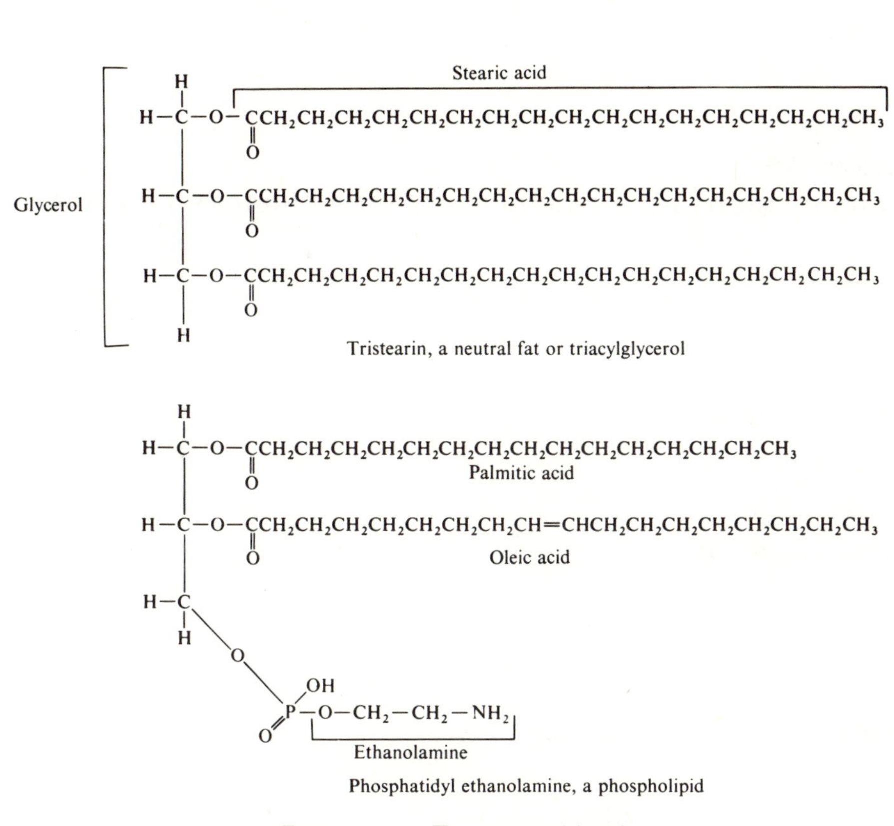

Figure 7–7. Two typical lipids.

building blocks are glycerol and three molecules of long-chain fatty acids (Fig. 7–7). Somewhat more complex and functionally more important are the *phospholipids*, which are important structural elements in cell membranes. Figure 7–7 also shows the structure of a typical phospholipid molecule; its building blocks are two fatty acid molecules (palmitic acid and oleic acid), and one molecule each of phosphate and the two alcohols *glycerol* and *ethanolamine*. This lipid is *phosphatidyl ethanolamine*, one of the most abundant phospholipids in animal tissues. We note that phosphatidyl ethanolamine contains two aliphatic ester linkages between the fatty acids and glycerol, and a phosphodiester "bridge" between two different alcohols, each of which requires a specific type of enzyme reaction pattern in its biosynthesis.

The first stage in the biosynthesis of phosphatidyl ethanolamine consists of the activation of the fatty acids and glycerol in ATP-linked reactions. This is then followed by assembly of the activated building blocks to form the complete lipid molecule. The structures of the activated building blocks are given in Fig. 7–8 and the reaction equations for their formation are given in

Palmitoyl CoA

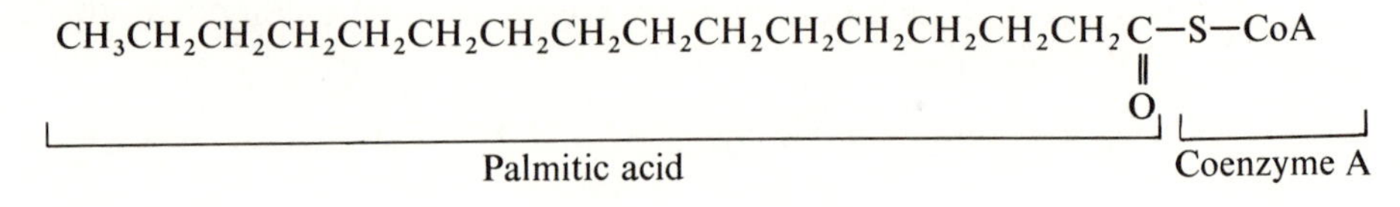

Glycerol 3-phosphate

```
 CH2OH
 |
HCOH      OH
 |        |
 CH2—O—P—OH
          ||
          O
```

Cytidine diphosphate diacylglycerol

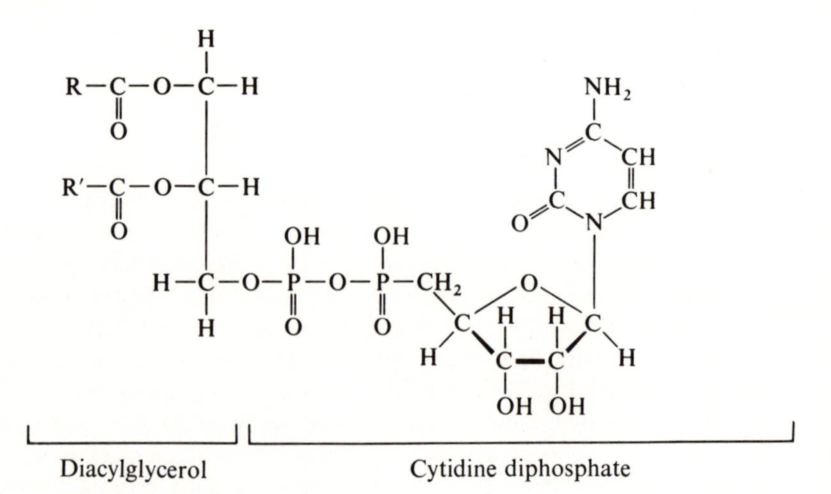

Figure 7–8. Activated building blocks in biosynthesis of phospholipids.

TABLE 7–3. Enzymatic Steps in the Biosynthesis of Phosphatidyl Ethanolamine

Activation of fatty acids

RCOOH + ATP + CoA—SH ⟶ R—CO—SCoA + AMP + pyrophosphate
R′COOH + ATP + CoA—SH ⟶ R′CO—SCoA + AMP + pyrophosphate
2AMP + 2ATP ⟶ 4ADP
2 pyrophosphate + $2H_2O$ ⟶ 4 phosphate

Activation of glycerol

glycerol + ATP ⟶ glycerol 3-phosphate + ADP

Formation of CDP-diacylglycerol

RCO—SCoA + glycerol 3-phosphate ⟶ monoacylglycerol 3-phosphate + CoA—SH
R′CO—SCoA + monoacylglycerol 3-phosphate ⟶ diacylglycerol 3-phosphate + CoASH
diacylglycerol 3-phosphate + CTP ⟶ CDP—diacylglycerol + pyrophosphate
pyrophosphate + H_2O ⟶ 2 phosphate

Attachment of ethanolamine

CDP—diacylglycerol + serine ⟶ phosphatidyl serine + CMP
phosphatidyl serine ⟶ phosphatidyl ethanolamine + CO_2
ATP + CMP ⟶ ADP + CDP
ATP + CDP ⟶ ADP + CTP
Sum: RCOOH + R′COOH + glycerol + serine + 7ATP ⟶ phosphatidyl ethanolamine + 7ADP + 6 phosphate + CO_2

Table 7–3. Altogether there are 15 enzymatic steps. The fatty acids are activated as esters of coenzyme A and the glycerol as the phosphate ester in ATP-dependent reactions. These activated building blocks are then assembled to yield phosphatidic acid, or diacylglycerol 3-phosphate. This molecule is then activated by reaction with CTP to yield the nucleotide derivative *cytidine diphosphate diacylglycerol.* Just as uridine nucleotides are specific carriers of monosaccharide units in polysaccharide synthesis, cytidine nucleotides are specific carriers of phosphatidic acid groups in the synthesis of phospholipids.

Two other points may be noted with regard to the energetics of these reactions. The first two equations in Table 7–3 describe the activation of fatty acids by a *pyrophosphate cleavage* of ATP; the pyrophosphate formed in each step must later be hydrolyzed to orthophosphate. Formation of each molecule of activated fatty acid (fatty acyl CoA) thus requires input of *two* high-energy phosphate bonds. The other point is that the adenosine 5′-monophosphate (AMP) formed in the activation of fatty acids must be rephosphorylated to

ATP in two steps. In the first step, AMP is phosphorylated to ADP at the expense of ATP by the enzyme *adenylate kinase*

$$\text{AMP} + \text{ATP} \rightleftharpoons \text{ADP} + \text{ADP}$$

The ADP so formed may then be phosphorylated to ATP directly during glycolysis or oxidative phosphorylation. The CMP formed from CTP must also be rephosphorylated to CTP; this process takes place in two steps, as is seen in Table 7–3.

If we now add up all the reactants and products of the 15 reaction steps required for synthesis of this phospholipid, which are catalyzed by a total of eleven different enzymes and cancel out those components appearing on both sides of the resulting equation, the sum will be the overall equation shown in Table 7–3. We see that seven molecules of ATP are ultimately split to ADP and phosphate during the synthesis of one molecule of phosphatidyl ethanolamine. If we now recall from Table 7–1 that a single *E. coli* cell can construct about 12,500 lipid molecules per second, it is evident that living cells are extremely busy in a chemical sense. Because there are 15 steps in the synthesis of each molecule of this phospholipid, approximately $15 \times 12{,}500$ or 187,500 chemical reactions must take place each second in an *E. coli* cell to produce its lipid molecules alone.

It may now be of some interest to make an approximation of the overall free energy change as a phosphatidyl ethanolamine molecule is constructed from its major building blocks. Exact figures are not available for each individual reaction step, but a reasonable approximation is possible for the overall reaction. The free energy input required for each of the two aliphatic ester linkages is about +3.5 kcal/mole, for the ester linkage between phosphate and ethanolamine, about +3.5 kcal/mole, and for the ester linkage between glycerol and phosphate, about +6.5 kcal/mole, or a total of about 17.0 kcal/mole of phosphatidyl ethanolamine constructed. This synthesis is achieved at the expense of seven moles of ATP, or a total of $7 \times 7.3 = 51.1$ cal/mole. The approximate efficiency of synthesis is thus about

$$(17.0/51.1)100 = 33\%$$

In most cells, there are perhaps ten or more different kinds of lipids, each of which has a specific set of building blocks, linked together in a specific way. Each requires for its biosynthesis a set of specific enzymes. Moreover, the various types of phospholipid are synthesized in specific molar ratios to each other. Thus the biosynthesis of the lipids can be an extremely elaborate and complex affair, even though these compounds are relatively small compared to polysaccharides.

7–10 ATP TURNOVER DURING ACTIVE BIOSYNTHESIS

The data in Table 7–1 include estimates of the total number of ATP molecules required per second for the biosynthesis of each major cell component, calculated on the basis of some reasonable approximations. It is assumed, that each phospholipid molecule requires about seven molecules of ATP, each polysaccharide molecule about 2000, each protein molecule about 1500, each RNA molecule about 6000, and each DNA molecule about 120,000,000. Thus about 2,500,000 molecules of ATP are broken down to ADP and phosphate per second in order to achieve the biosynthesis of these components of an *E. coli* cell.

Since a single *E. coli* cell contains about five million molecules of ATP, enough for only two seconds of biosynthetic work, ATP must be constantly regenerated from ADP and phosphate by oxidative phosphorylation. In the metabolic steady state of an *E. coli* cell the half-time of turnover of the terminal phosphate group of its ATP must therefore be in the neighborhood of one or two seconds. The steady-state concentration of ATP in the cell is thus the resultant of two relatively massive and opposing processes, the utilization of ATP and its regeneration. Whenever the steady-state level of ATP in the cell decreases because of the onset of a period of accelerated biosynthesis, or any other ATP-requiring activity, the rate of respiration and thus the rate of ATP production automatically speeds up through the action of the allosteric enzymes controlling the rates of glycolysis and respiration. Conversely, whenever the steady-state level of ATP increases, through a temporary decrease in the rate of its utilization for biosynthetic or other purposes, the rate of respiration and thus the rate of ATP resynthesis tends to slow down. At all times cells tend to maintain steady states, not only of ATP but of all cell components, such as lipids, polysaccharides, and even proteins, each of which undergoes constant metabolic turnover.

7–11 THE METABOLIC STEADY STATE AND ENTROPY PRODUCTION

Now we may ask an important question. What is the significance of the dynamic steady state of cell components, in which the rate of their synthesis is exactly balanced by the rate of their degradation?

Because living cells are open systems, and because they are not in thermodynamic equilibrium with their surroundings, their energy exchanges are most appropriately analyzed in terms of the principles of nonequilibrium thermodynamics, an extension of classical equilibrium thermodynamics discussed in Chapter 2. In the theory of nonequilibrium or open-system thermodynamics the steady state possesses the same meaning as the equilibrium state in equilibrium thermodynamics. In brief, the steady state is the most orderly, efficient, and economical state of an open system; indeed we may say that the steady state is a characteristic of all smoothly running machinery. This is because the rate of entropy production by an open system is at a minimum when the system is in a steady state. This important thermodynamic principle has been commented on in the following words:

> "This remarkable conclusion . . . sheds new light on "the wisdom of living organisms." Life is a constant struggle against the tendency to produce entropy by irreversible processes. The synthesis of large and information-rich macromolecules, the formation of intricately structured cells, the development of organization—all these are powerful anti-entropic forces. But since there is no possibility of escaping the entropic doom imposed on all natural phenomena under the Second Law of thermodynamics, living organisms choose the least evil —they produce entropy at a minimal rate by maintaining a steady state."[1]

[1] A. Katchalsky, "Non-equilibrium Thermodynamics," *Modern Science and Technology*, R. Colbern, ed., Van Nostrand, New York, 1965, p. 194.

8

BIOSYNTHESIS OF DNA, RNA, AND PROTEINS

In the preceding chapter we examined in some detail the biosynthesis of two of the major chemical components of living cells, the lipids and polysaccharides. Because they are relatively simple molecules we were able to focus on the thermodynamic principles and the enzymatic reaction patterns by which the biosynthesis of most cellular compounds is achieved.

As we examine the biosynthesis of the nucleic acids and the proteins we shall encounter an aspect of biosynthetic processes that we have not yet considered, namely, the incorporation and transfer of information. The nucleic acids are molecules adapted for the storage, replication, and transcription of genetic information, and the proteins serve to express it.

We shall see that the biosynthesis of the nucleic acids and proteins is much more complex than that of polysaccharides and lipids, not because the types of chemical bonds linking together their building blocks are particularly complicated or difficult to make, but because the building blocks must be inserted into the structure of these chainlike molecules in a precisely specified order or sequence. We shall see how the phosphate bond energy of ATP is employed to form the chemical linkages between the building blocks of nucleic acids and proteins and to ensure the precision with which their characteristic sequences are assembled.

8-1 ELEMENTS IN THE FLOW OF GENETIC INFORMATION

First, for orientation, we shall outline the major stages in the flow of genetic information in the cell. There are three major kinds of molecular elements in genetic information transfer, deoxyribonucleic acid (DNA), ribonucleic acid (RNA), and proteins. These elements interact in such a way that the flow of genetic information under usual circumstances proceeds from DNA of the parent cell to DNA of the daughter cells, and, within any given cell, from DNA to RNA to protein, according to the scheme

parent cell	DNA	⟶	RNA	⟶	protein
	↓				
first generation	DNA	⟶	RNA	⟶	protein
	↓				
second generation	DNA	⟶	RNA	⟶	protein
	↓				

This set of relationships is often referred to as the *central dogma* of molecular biology. It was first propounded as a working hypothesis in the 1950's by Crick and it has since been verified by over a decade of penetrating experimental investigations.

Now let us examine this flowsheet more closely and define some important terms.

Deoxyribonucleic acid (DNA), which is largely located in the nucleus, is the "master" informational molecule of the cell; it contains all the genetic information required for the precise replication of cells. It is a very long chainlike molecule containing a characteristic sequence of four different kinds of building blocks, the deoxyribonucleotides. When the cell divides, the parental DNA strands are used as templates for the enzymatic synthesis of daughter strands of DNA from simple precursors, in such a way that each of the daughter cells receives DNA molecules identical in sequence to the parental DNA. This process is called *replication*, and it represents the means for the precise transmission of genetic information from each generation to the next.

The genetic information in the DNA molecule is ultimately used to specify the characteristic sequence of building blocks of proteins during their synthesis. However, the DNA itself is not used directly as the working template for this process; rather it should be regarded as a "master tape," that is kept locked in a safe, that is, the nucleus. From this master tape working tapes are from time to time transcribed for use as templates in the synthesis of cell proteins. These working tapes, which carry the genetic message from the nucleus to the ribosomes, are specialized RNA molecules called messenger RNA's. Messenger RNA's are also long chainlike molecules. They contain four different building blocks, the ribonucleotides, in a sequence that is precisely complementary to that of the DNA strand from which they were transcribed.

The synthesis of messenger RNA molecules is known as *transcription*; it occurs in the nucleus, directly alongside the DNA strand being transcribed. Clearly, transcription must proceed with just as great molecular precision as the replication of DNA if each cell is to have a characteristic phenotype or set of heritable characteristics.

Finally, in the last great stage of genetic information transfer, the sequences of coding elements inherent in the structure of messenger RNA molecules are utilized as templates in construction of the many different kinds of protein molecules, that ultimately determine the characteristic size, shape, structure, and activities of each type of cell. Proteins, like nucleic acids, are chainlike molecules consisting of recurring covalently joined units, namely twenty different simple building blocks, the amino acids. During the synthesis of proteins the 4-symbol language of DNA and its corresponding messenger RNA must be translated into the 20-letter language of protein structure. Hence this stage of genetic information transfer is spoken of as *translation*.

Now let us examine the molecular structure of each of the major elements of the genetic system and how they are synthesized from simpler precursors.

8–2 THE STRUCTURE OF DNA

Four different deoxyribonucleotides constitute the building blocks of DNA (Fig. 8–1). They are identical in structure except for the nitrogenous base components; they differ from the ribonucleotides discussed in Chapter 7 in that they lack a hydroxyl group at carbon atom 2 of the ribose. In the DNA molecule these nucleotide units are arranged in long chains (called polynucleotides) through phosphodiester bridges, which link successive nucleotides between the 5′-carbon atom of the deoxyribose of one nucleotide with the 3′-carbon atom of the next, as is shown in Fig. 8–2. Thus the covalent backbone of DNA consists of alternating phosphate and deoxyribose units, from which extend side chains consisting of the different bases in a characteristic sequence. The genetic message that the DNA molecule carries is imparted by the specific sequence of the four different bases along the chain; a sample sequence would be A—T—G—T—C—A—G—C—T. It is obvious that an enormous number of different sequences is potentially possible, particularly since most DNA molecules are very long, with millions of nucleotide units.

In 1953 Watson and Crick deduced from x-ray diffraction analysis of native DNA that it exists under biological conditions as a double-helical structure, in which two strands of DNA are intertwined helically around each other in such a way that the two molecules run in opposite directions and the bases of the two strands fit each other in a complementary fashion (Fig. 8–3). The diameter of the DNA double helix is 20 Å and the axial

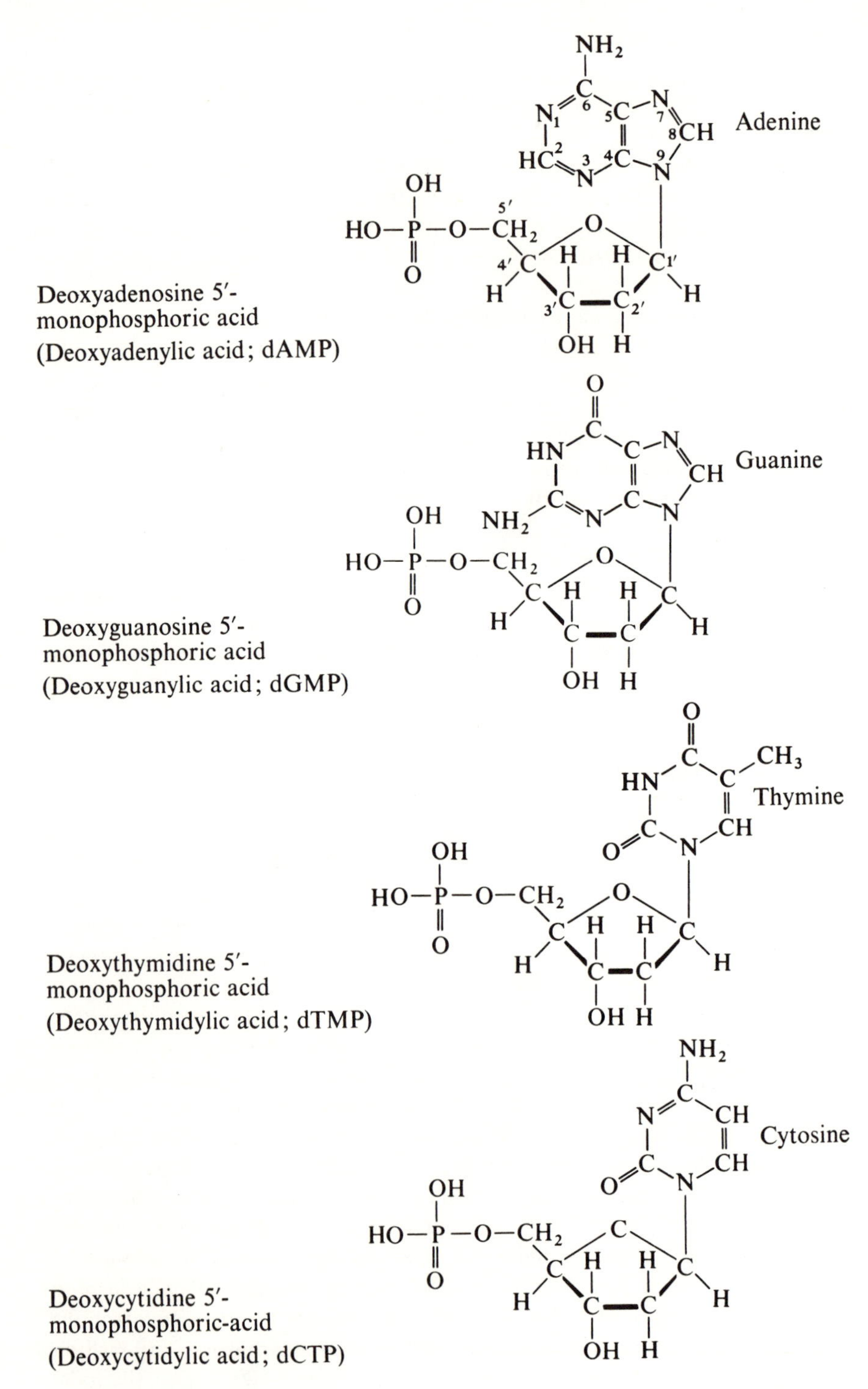

Figure 8–1. The deoxyribonucleotide building blocks of DNA.

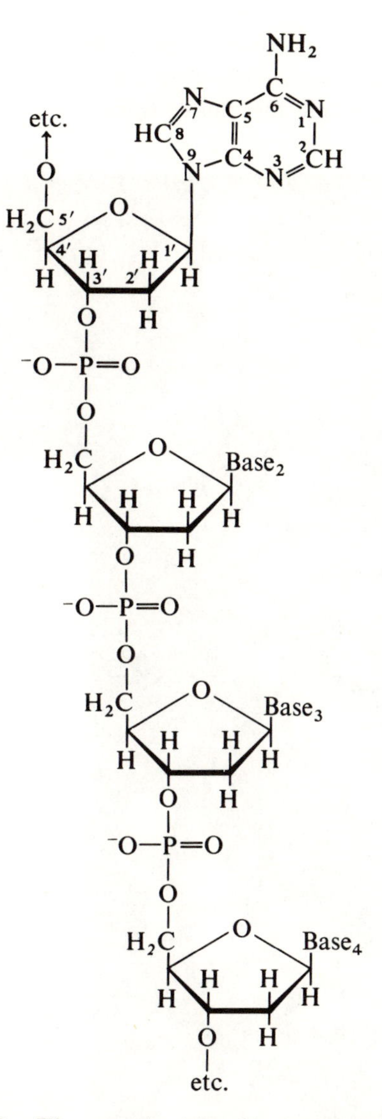

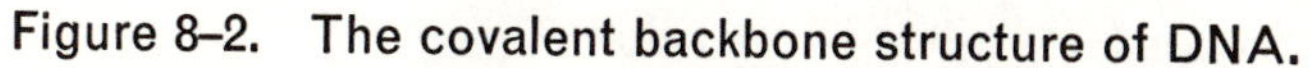
Figure 8–2. The covalent backbone structure of DNA.

distance between adjacent base pairs is 3.4 Å; in one complete turn of the double helix there are ten base pairs. The double helix has a major and a minor groove, as is seen in the model (Fig. 8–3). The entire double-helical molecule is held together laterally by hydrogen bonds between the uniquely fitting pairs of bases (see Fig. 8–4) and vertically by hydrophobic interactions between the stacked bases. The association of the flat, water-insoluble bases in stacks is a consequence of the tendency of the surrounding water molecules to seek their most random or entropy-rich configuration, which forces the DNA strands to seek that steric conformation in which the base molecules are hidden from water, within the double helix.

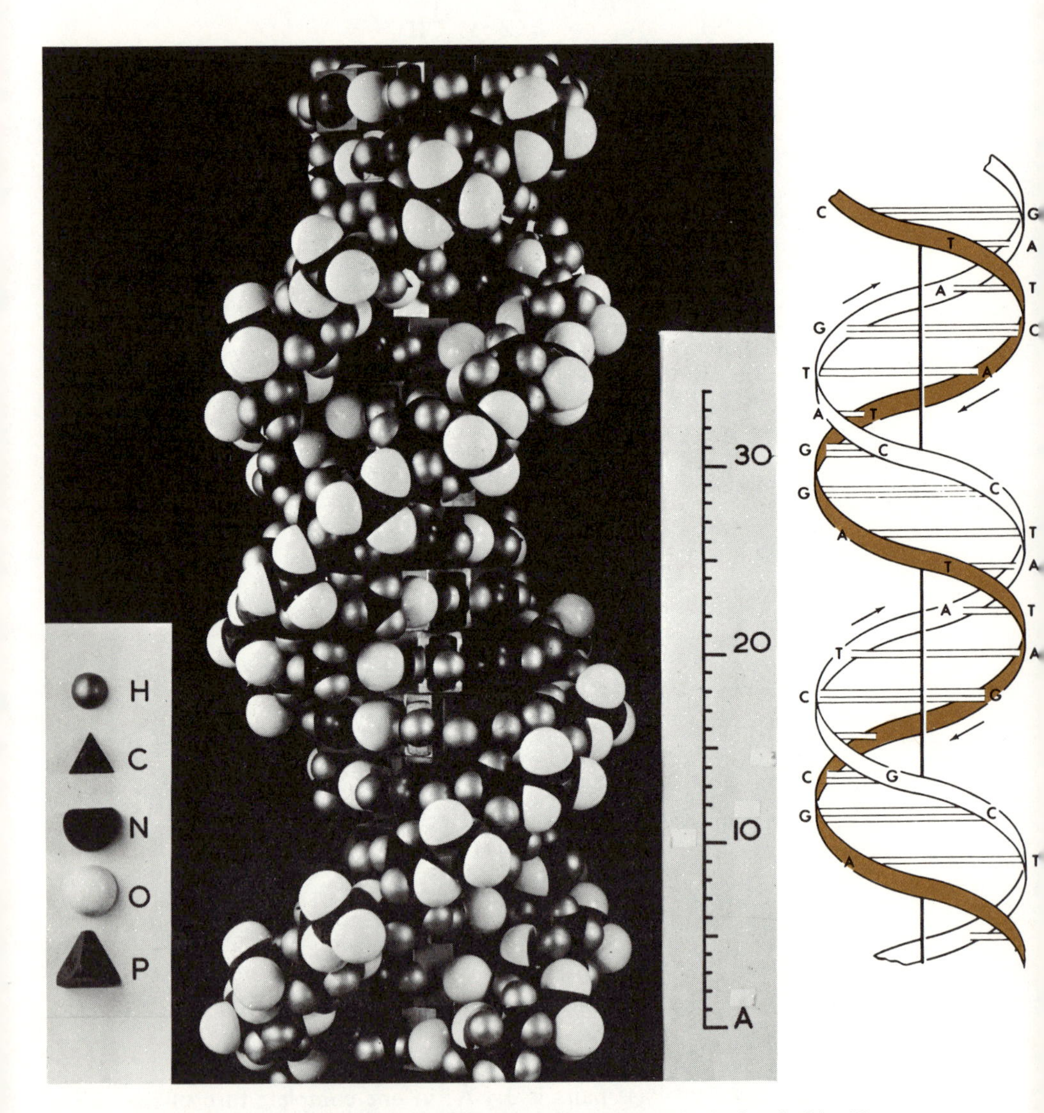

Figure 8–3. Molecular model of the DNA double helix (left). (Courtesy of M. H. F. Wilkins, Kings College, London). At the right is a schematic representation showing the pairing of bases (A=T and G≡C) within the helix. The opposite polarity of the strands is shown by the arrows.

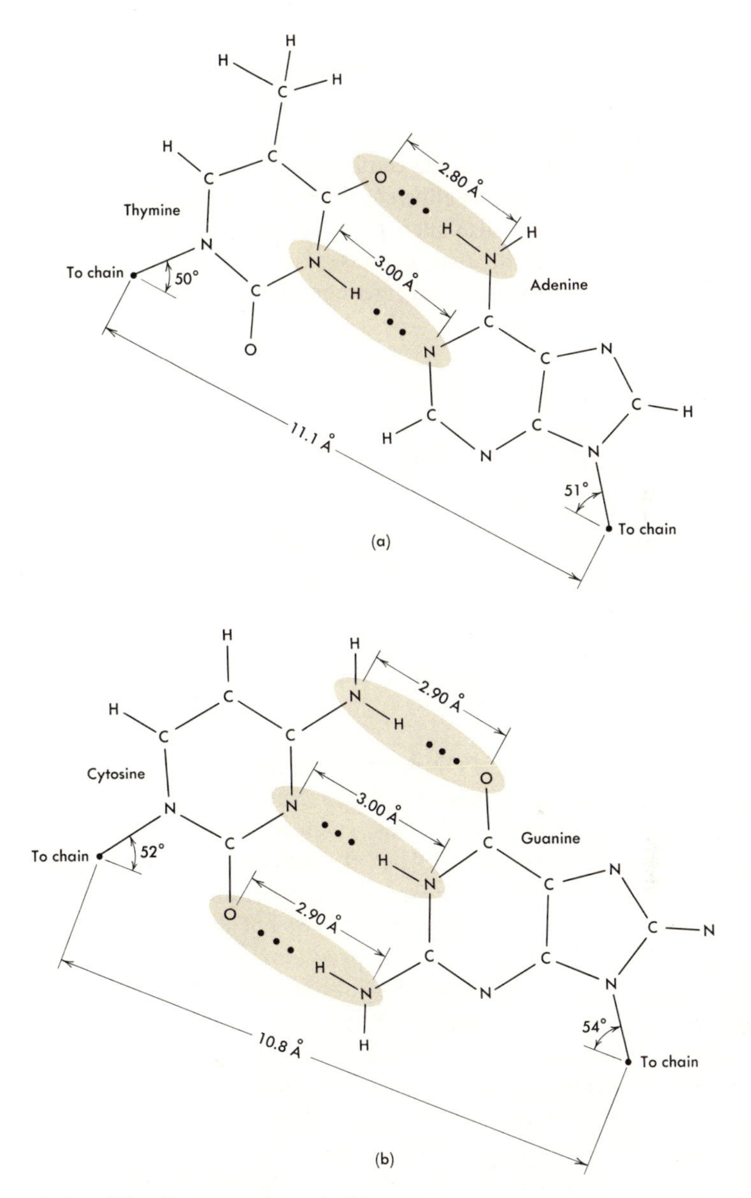

Figure 8–4. The base pairs of DNA, showing the complementary fit between (a) adenine and thymine and between (b) guanine and cytosine. Because there are three hydrogen bonds between guanine and cytosine this base pair is more stable than the adenine-thymine pair.

Both strands of the DNA double helix contain exactly the same information because of the unique and complementary pairing of bases. The complementary base pairs are adenine and thymine (or A—T) and cytosine and guanine (or C—G). As shown in Fig. 8–4, these pairs of bases can hydrogen bond with each other through their complementarity of structure; the resulting pairs fit exactly within the central axial "hole" of the double helix. Other pairings of bases are not allowed, either because stable hydrogen bonding is not possible, or because such pairs do not fit within the helix. It is important to note that the two strands of double-helical DNA are not identical in sequence; rather, they are complementary, as is shown in a simple diagram

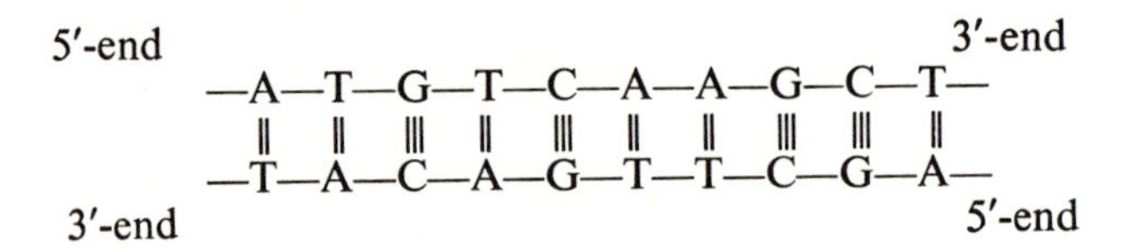

The designations "5′-end" and "3′-end" indicate the ends of the chain bearing a free hydroxyl group on the terminal deoxyribose unit at the 5′ and 3′ positions respectively.

8–3 THE STRUCTURE OF CHROMOSOMES, GENES, AND CODONS

Now let us consider the size of DNA molecules in terms of the basic functional units of genetic inheritance, namely, the chromosome and the gene. Here it is useful to contrast prokaryotic cells, such as the bacterium *E. coli*, and eukaryotic cells. In prokaryotic cells all the genetic information is present in a single chromosome, which consists of one enormous double-helical DNA molecule. In *E. coli* the chromosome has a weight of 2,800,000,000 daltons (one dalton equals the weight of a hydrogen atom) and the complete double helix is about 1.2 mm long. It contains about 4,200,000 base pairs along its length. However, the chromosome of *E. coli* is not a single long structure with two ends but rather an endless closed loop, covalently linked throughout. In fact, it appears likely that the chromosomes of all cells normally have such an endless or circular structure. In the eukaryotic cells of higher organisms, the DNA is divided among several or many chromosomes, each of which may be a single, very large DNA molecule of endless circular structure. The higher the organism in the phylogenetic scale, and thus the more genetic information required to specify its replication, the more DNA its cells contain.

Each chromosome contains many genes; the single chromosome of *E. coli*, for example, contains 3000 or more genes. A gene is defined in modern molecular terms as that segment of a chromosome coding for the synthesis of a single polypeptide chain of a protein molecule or for single molecules of certain types of RNA—namely, transfer RNA and ribosomal RNA, whose nature and role we shall examine later. The size of genes varies considerably, since the size of the polypeptide chains of different proteins varies. Most genes are about 300 to 900 nucleotide units long, but some have as few as 75 nucleotides and others as many as 5000. We may thus visualize the long DNA chain of the chromosome as being functionally subdivided into segments, each of which codes for a specific polypeptide chain; such segments are also called *cistrons*. Moreover, genes have specific locations along the chromosome whose sequence can be deduced by genetic mapping methods.

DNA molecules contain still smaller functional units called *codons*. Codons, as the name implies, are the code words of genes; each codes for a single amino acid residue in the polypeptide chain that is specified by the gene. From both genetic and biochemical experiments it is now known that each codon consists of three successive nucleotide units in the DNA chain, sometimes called a coding triplet. Since there are four different bases in DNA, and since these bases can be arranged in a total of 64 groups of three, all having different base sequences, there are 64 different possible codons in the "language" of genes. Codons are arranged in a sequence along the gene that corresponds to the sequence of amino acid units in proteins, without the use of "punctuation" nucleotides. Thus a gene having a total of 450 nucleotide residues contains $450/3 = 150$ codons and could thus specify the amino acid sequence of a polypeptide chain having 150 residues.

Coding of RNA by DNA involves a one-for-one correspondence of nucleotides through the base-pairing principle, such as is seen in the DNA double helix.

8-4 REPLICATION OF DNA

The Watson-Crick model for the replication of DNA predicted that each of the two complementary strands of the parental chromosome are replicated to yield two daughter double helixes (also called duplexes), of which one strand was contributed by the parent and the other synthesized *de novo* from mononucleotide precursors. This principle, called *semiconservative replication*, has been experimentally verified (Fig. 8-5). Moreover, it has also been established that when the circular chromosome of *E. coli* is replicated, there is usually a single growing point at which the two strands of the parental DNA are parted and the two new strands are formed. This growing point normally moves around the entire circular chromosome once during each cell division time of the bacterium, which is anywhere from 20 to 60 minutes depending on the culture ingredients and other conditions. The end result is two new circular double-helical chromosomes, which pass to the two

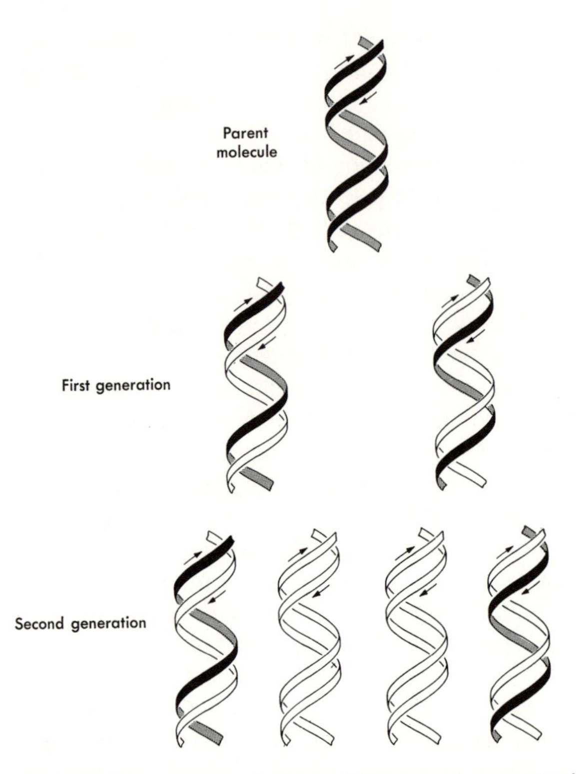

Figure 8–5. Schematic representation of the semiconservative replication of DNA. The white strands in the first and second generations are the newly formed strands.

daughter cells during division. The molecular mechanism of DNA replication must therefore account for: (a) how each mononucleotide unit is inserted into the new chain, (b) how a precise complementary sequence of bases is produced in the new strand, and (c) how both strands are made simultaneously even though the two strands of the parental DNA run in opposite directions.

Kornberg and his colleagues have discovered an enzyme, DNA polymerase, which is believed to be the major enzyme concerned in DNA synthesis; it is found in all cells. This enzyme, which has been isolated in pure form from *E. coli*, is capable of building the linkage between successive mononucleotide units of DNA in a reaction that requires as precursors the 5′-triphosphates of each of the four deoxyribonucleosides, namely, dATP, dGTP, dTTP, and dCTP; dATP is shown in Fig. 8–6. Mg^{2+} ions are also required. The enzyme causes the sequential formation of new phosphodiester linkages by splitting out the terminal two phosphate groups of the precursor triphosphate to yield inorganic pyrophosphate; simultaneously, a covalent linkage is formed

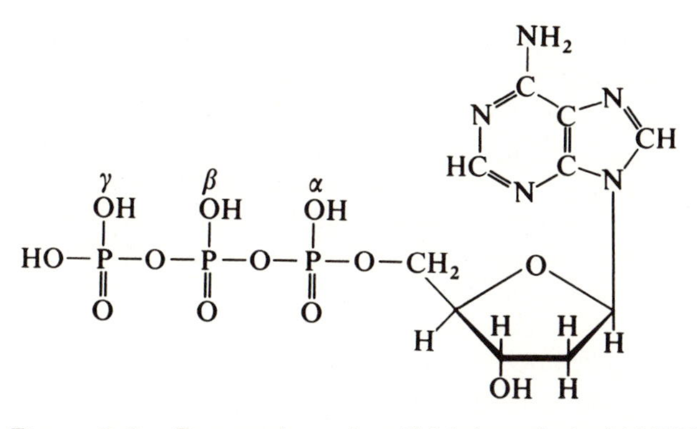

Figure 8–6. Deoxyadenosine 5′-triphosphate (dATP).

between the remaining phosphate group and the free 3′-hydroxyl group of the terminal nucleotide residue on the chain being built. The DNA chain is thus built in the 5′→3′ direction; new units are attached only at the end having a free 3′-hydroxyl group. We may now write the equation for the introduction of each nucleotide unit as

$$\text{dNTP} \xrightleftharpoons{\text{Mg}^{2+}} (\text{dNMP})_{\text{DNA}} + \text{PP}_i$$

in which dNTP denotes a deoxyribonucleoside triphosphate, dNMP denotes a mononucleotide unit of DNA, and PP_i denotes inorganic pyrophosphate. This reaction is reversible as written. However, the pyrophosphate formed undergoes subsequent enzymatic hydrolysis catalyzed by pyrophosphatase

$$\text{PP}_i + \text{H}_2\text{O} \rightarrow 2\text{P}_i$$

in a reaction that has a large negative $\Delta G^{0\prime}$ value of about -7.3 kcal. The overall reaction by which the mononucleotide unit is inserted is therefore

$$\text{dNTP} + \text{H}_2\text{O} \rightleftharpoons (\text{dNMP})_{\text{DNA}} + 2\text{P}_i$$

This reaction is strongly exergonic, with an overall $\Delta G^{0\prime}$ of at least -7.3 kcal. Thus two high-energy phosphate bonds are ultimately required to make each internucleotide linkage of DNA.

However, DNA polymerase requires another component not specified by this equation before it can introduce new nucleotide units to the end of the DNA chain, namely, a strand of preformed or parental DNA that serves as a template. The new strand is built alongside the template strand in such a manner that the new chain becomes complementary to the template. In other words, wherever an adenine residue appears in the template strand, the complementary thymine nucleotide is inserted into the new strand, and vice versa. Similarly, wherever a guanine appears in the template, a cytosine nucleotide will be introduced, and vice versa. Moreover, it has been established that the end product of the action of DNA polymerase is not single-stranded but double-helical DNA, in which one strand is the unmodified template and the other the newly formed strand. There are therefore two major factors that cause the synthesis of a new DNA strand to be an essentially irreversible process. The first is the hydrolysis of pyrophosphate and the second is the thermodynamic "pull" exerted by the tendency of the newly formed strand to enter into formation of the stable double helix with the template strand. Thus the faithful replication of DNA is guaranteed by the operation of fundamental thermodynamic principles.

Recent research has shown that the new strand of DNA is not made in one long unbroken stretch. Rather, it is made in short segments of less than 1000 nucleotide residues, which are then connected by another enzyme, DNA ligase, which functions to "ligate" or tie together the short stretches synthesized by DNA polymerase. This process may at first appear unnecessarily complex, but it has a very specific purpose, since it is believed to make possible the replication of *both* strands of the parental DNA simultaneously, as is seen in Fig. 8–7. The DNA polymerase first makes a segment of new DNA complementary to one strand of the parental DNA and then "jumps" to the other parental strand and replicates it, going backwards (but always in the $5' \rightarrow 3'$ direction). After this hairpin-like loop is completed, it is cleaved enzymatically at the point where the parental DNA strands are parted, and its proximal ends attached to the ends of the previously synthesized stretches of the new DNA strands. In this way the DNA polymerase, which works only in the $5' \rightarrow 3'$ direction, is believed to replicate both strands of the parental DNA, even though they run in opposite directions.

There is now significant evidence that *E. coli* cells contain two types of DNA polymerase. One is concerned in replication of entire DNA strands, whereas the other may function to repair breaks or other types of damage in short segments of DNA strands.

Recently it has been discovered that DNA may, under special circumstances, be synthesized from an RNA template. There are several animal viruses containing RNA, but no DNA, which have been known to cause normal mammalian cells to undergo transformation into cancer cells; this transformation is heritable. These RNA viruses have been found to contain a special type of DNA polymerase. This enzyme requires deoxyribonucleoside 5′-triphosphates as precursors but does not utilize DNA as template strand. Rather it utilizes as template the RNA present in the virus. The newly formed DNA transcribed from the RNA appears to become incorporated into the host mammalian cell DNA. When the latter is replicated the new segment of DNA is also replicated, thus causing the progeny cells to contain genes introduced by the virus. This discovery not only suggests how the malignant transformation of normal cells may take place but it also is the first reported instance of the "backward" flow of genetic information, from RNA to DNA.

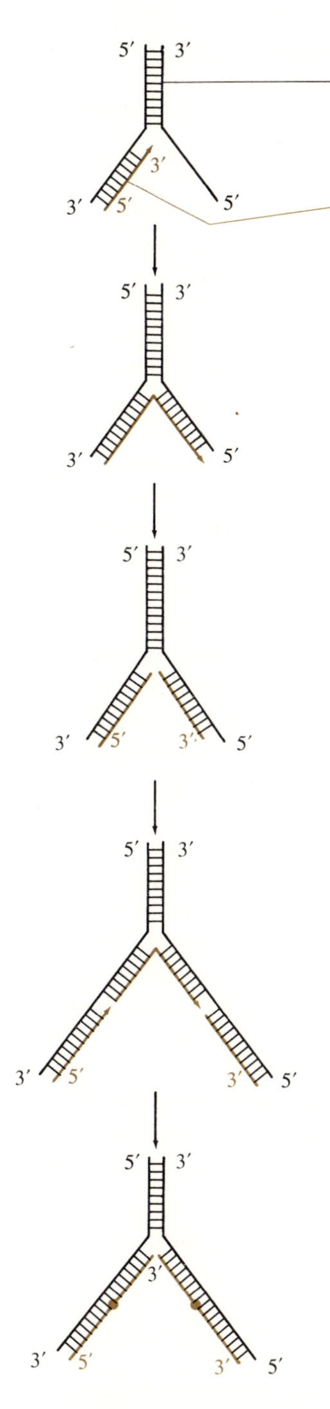

Figure 8–7. Replication of both strands of DNA by DNA polymerase and DNA ligase.

8–5 MESSENGER RNA AND THE TRANSCRIPTION PROCESS

Messenger RNA is one of three major types of ribonucleic acid in cells: messenger RNA (mRNA), transfer RNA (tRNA), and ribosomal RNA (rRNA). Messenger RNA is the least abundant, making up only a small per cent of the total cell RNA. However, it usually has a relatively high turnover rate, that is, it is rapidly synthesized and rapidly degraded. All three types of RNA's have a covalent backbone structure like that of DNA, except for two significant differences. The first is that the mononucleotide units are ribonucleotides and contain the 5-carbon sugar D-ribose instead of the 2-deoxy-D-ribose found in DNA. Secondly, the four characteristic major bases of RNA are adenine, guanine, cytosine, and uracil; uracil replaces the thymine found in DNA. The internucleotide linkages are 3′, 5′-phosphodiester bonds, as in DNA.

Messenger RNA differs in other respects from DNA. It is a single-stranded molecule and does not tend to form a double helix with another strand of RNA. Messenger RNA molecules vary considerably in size. The shortest mRNA's have about 300 nucleotide units; the longest may have as many as 3000 or more bases. Messenger RNA molecules correspond in length and are complementary in base sequence to a single gene or to a group of related genes present in adjacent sequence in the DNA; such a group of genes is called an *operon.*

The transcription of the genetic message embodied in a specific base sequence in the DNA molecule proceeds by an enzymatic process that is analogous to that taking place in the replication of DNA chains. A specific enzyme present in the nucleus, called *DNA-directed RNA polymerase,* catalyzes the formation of the covalent backbone of mRNA from the four ribonucleoside 5′-triphosphate precursors, namely, ATP, GTP, UTP, and CTP (see Chapter 7 for their structures). For this reaction a strand of DNA is required as the template; without the template, no RNA synthesis occurs. A molecule of pyrophosphate is lost from each mononucleotide 5′-triphosphate unit acted on; the pyrophosphate formed then undergoes secondary hydrolysis into two molecules of inorganic phosphate. These reactions may be written as follows

$$\mathrm{NTP} \rightleftharpoons (\mathrm{NMP})_{\mathrm{RNA}} + \mathrm{PP_i}$$

$$\mathrm{PP_i} + \mathrm{H_2O} \longrightarrow 2\mathrm{P_i}$$

$$\text{Sum: } \mathrm{NTP} + \mathrm{H_2O} \longrightarrow (\mathrm{NMP})_{\mathrm{RNA}} + 2\mathrm{P_i}$$

Two high-energy phosphate bonds are thus required to generate each internucleotide linkage. At each step, that nucleoside triphosphate is selected that can fit the corresponding base in the template DNA in a sterically complementary way. G and C are always complementary; however, when A is the base in the template DNA, then U is the base in the mRNA, and when T is the base in DNA then A is the base in mRNA. Although the RNA polymerase requires a double-stranded DNA molecule as template, only one strand of it is actually transcribed. The single-stranded mRNA forms a double-helical structure with the template strand of the DNA in an intermediate stage of the transcription reaction. This is called a *DNA-RNA hybrid.* The tendency for this hybrid to stabilize itself, through both hydrogen bonding between the complementary bases and the hydrophobic attraction between the stacked bases, exerts a further thermodynamic pull, in addition to that yielded by hydrolysis of pyrophosphate, which guarantees the exactness of the transcription process.

The newly formed messenger RNA ultimately "peels" away from the template DNA by a mechanism that is not yet understood; it then leaves the nucleus and is transported to the ribosomes in the cytoplasm. Much evidence suggests that there are special signals for RNA polymerase to "start" and "stop," since the different genes on a given chromosome may be transcribed at different times and at different rates.

8-6 THE COVALENT STRUCTURE OF PROTEINS

The last stage in the flow of genetic information is the translation of the genetic message carried by messenger RNA to form the proteins, which serve as the ultimate effectors or instruments for expression of genetic information. We will now examine the covalent structure of proteins and how it relates to their many different biological functions and their distinctive species specificity.

Proteins are large molecules consisting of recurring building blocks, the amino acids, which are covalently linked into long chains called polypeptides. There are twenty different amino acids found in proteins, regardless of the species of origin; their names, structures, and symbols are shown in Fig. 8-8. All of them have in common an amino group and a carboxyl group attached to the α-carbon atom; however, they differ in having distinctive side chains, usually called R groups, as is shown in the following generalized structural formula:

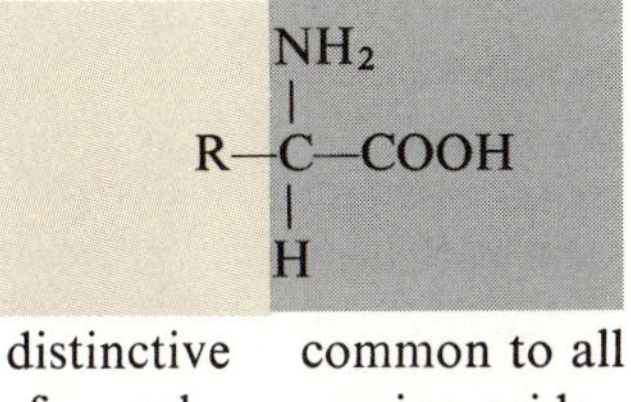

distinctive for each amino acid — common to all amino acids

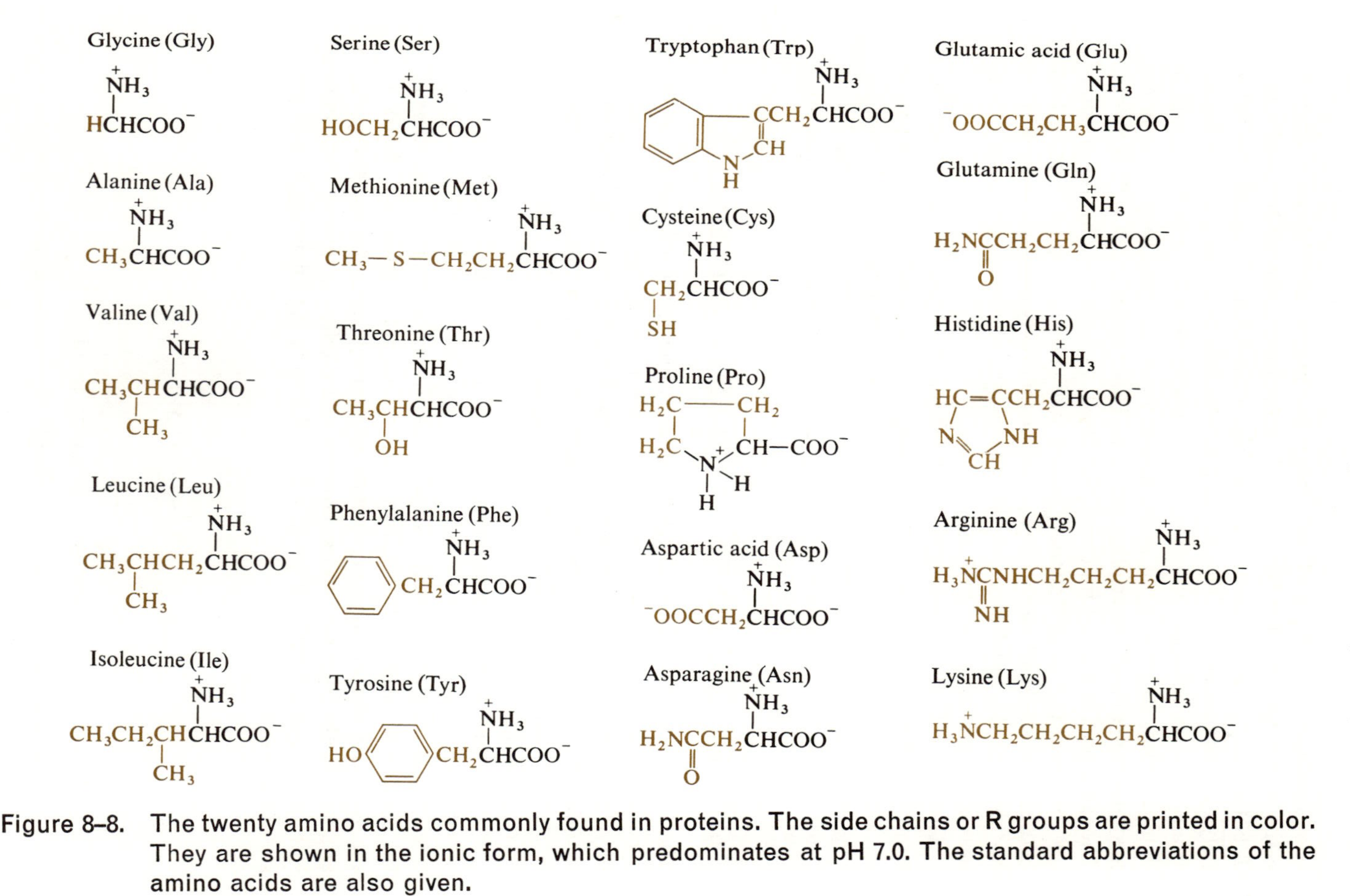

Figure 8–8. The twenty amino acids commonly found in proteins. The side chains or R groups are printed in color. They are shown in the ionic form, which predominates at pH 7.0. The standard abbreviations of the amino acids are also given.

As can be seen in Fig. 8–8, the R groups of the various amino acids differ in size, shape, and polarity. Some are highly polar and carry an electrical charge at the pH of the cell, such as the R groups of glutamic acid and lysine; such R groups are water soluble or hydrophilic (water-liking). However, the R groups of other amino acids, such as isoleucine and phenylalanine, are non-polar and thus are oily in nature or hydrophobic (water-hating).

Two amino acids may be covalently linked by the peptide bond, which joins the α-amino group of one and the α-carboxyl group of the other. When two amino acids are so joined, the product is a dipeptide; if further amino acids are added at either end through peptide linkages, we will have successively tripeptides, tetrapeptides, pentapeptides, and, in the case of very long chains, polypeptides (Fig. 8–9). Clearly, from the twenty different amino acids it is possible to construct many different polypeptides varying in the relative numbers of different amino acids and in their sequence. Moreover, the longer the polypeptide chain, the greater the number of possible amino acid sequences.

Most polypeptide chains of proteins contain between 100 to 300 amino acid residues, corresponding to molecular weights of 12,000 to 36,000. Small protein molecules, such as the enzymes ribonuclease and cytochrome *c* and the transport protein myoglobin, which have molecular weights of about 12,600, 13,400 and 16,900, respectively, contain but a single polypeptide chain. However, most of the larger protein molecules have more than one polypeptide chain; for example, hemoglobin, the oxygen-carrying protein of the red blood cell, contains four polypeptide chains, each having about 150 amino acid residues. The molecular weight of hemoglobin is about 64,500.

Each cell contains hundreds if not thousands of different proteins, all of which differ in their amino acid sequence. Each of these proteins has some distinctive and characteristic function in the cell. Among the most important proteins are the enzymes, of which over 1000 different catalytic types are known. Each of these has a characteristic molecular weight and a genetically determined amino acid sequence. Other types of proteins are specialized to serve structural functions, such as the collagen of connective tissue, the sheath proteins of viruses, the structural proteins of cell membranes, and the contractile proteins of muscle.

There is another way in which protein molecules differ from each other. The proteins present in each species of organism are characteristic of that species and can be chemically distinguished from homologous proteins of other species. Homologous proteins are those having identical functions; the enzyme cytochrome *c*, for example, is present in and catalyzes the same oxidation-reduction reaction in the cells of vertebrates, insects, bacteria, yeasts, fungi, and higher plant cells. Moreover, the cytochrome *c* of all these organisms are very similar chemically. Nevertheless, they can be distinguished

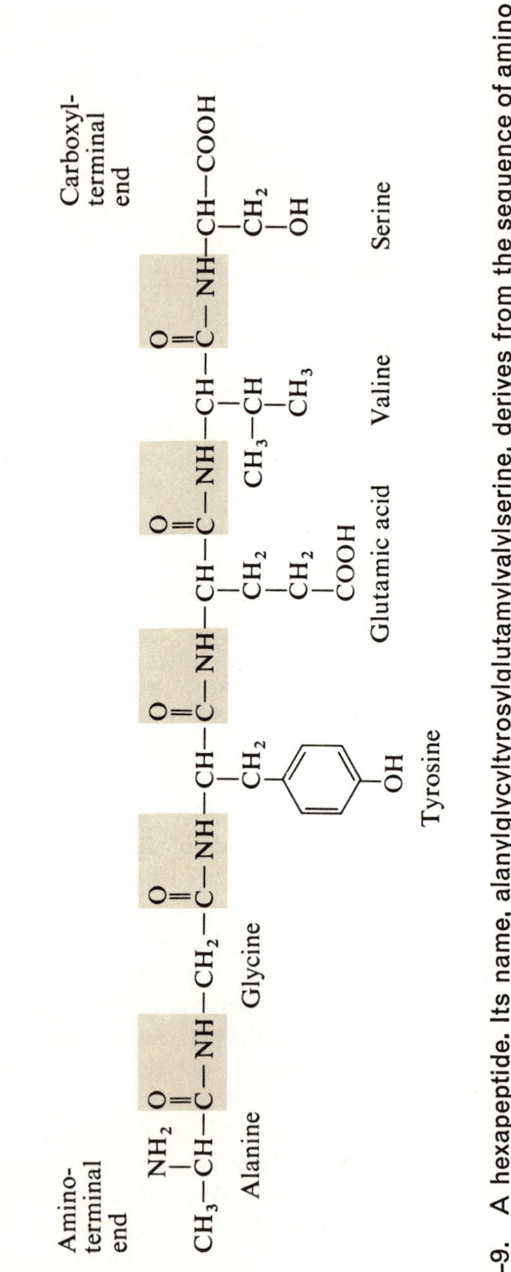

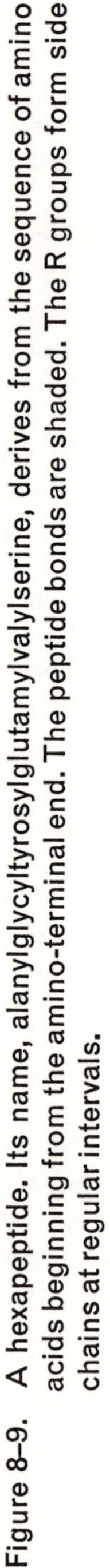

Figure 8–9. A hexapeptide. Its name, alanylglycyltyrosylglutamylvalylserine, derives from the sequence of amino acids beginning from the amino-terminal end. The peptide bonds are shaded. The R groups form side chains at regular intervals.

from each other since they differ to a greater or lesser extent in their amino acid sequence.

The complete amino acid sequence of many proteins is now known as the result of very painstaking analytical work. Table 8–1 shows as an example

TABLE 8–1. The Amino Acid Sequence of Sperm Whale Myoglobin. The amino acid symbols employed are given in Fig. 8–8

Amino end			Val	· Leu	· Ser	· Glu	· Gly	· Glu	· Trp	· Gln	· Leu	· Val	· Leu	· His
			1	2	3	4	5	6	7	8	9	10	11	12
Val	· Trp	· Ala	· Lys	· Val	· Glu	· Ala	· Asp	· Val	· Ala	· Gly	· His	· Gly	· Gln	· Asp
13	14	15	16	17	18	19	20	21	22	23	24	25	26	27
Ile	· Leu	· Ile	· Arg	· Leu	· Phe	· Lys	· Ser	· His	· Pro	· Glu	· Thr	· Leu	· Glu	· Lys
28	29	30	31	32	33	34	35	36	37	38	39	40	41	42
Phe	· Asp	· Arg	· Phe	· Lys	· His	· Leu	· Lys	· Thr	· Glu	· Ala	· Glu	· Met	· Lys	· Ala
43	44	45	46	47	48	49	50	51	52	53	54	55	56	57
Ser	· Glu	· Asp	· Leu	· Lys	· Lys	· His	· Gly	· Val	· Thr	· Val	· Leu	· Thr	· Ala	· Leu
58	59	60	61	62	63	64	65	66	67	68	69	70	71	72
Gly	· Ala	· Ile	· Leu	· Lys	· Lys	· Lys	· Gly	· His	· His	· Glu	· Ala	· Glu	· Leu	· Lys
73	74	75	76	77	78	79	80	81	82	83	84	85	86	87
Pro	· Leu	· Ala	· Gln	· Ser	· His	· Ala	· Thr	· Lys	· His	· Lys	· Ile	· Pro	· Ile	· Lys
88	89	90	91	92	93	94	95	96	97	98	99	100	101	102
Tyr	· Leu	· Glu	· Phe	· Ile	· Ser	· Glu	· Ala	· Ile	· Ile	· His	· Val	· Leu	· His	· Ser
103	104	105	106	107	108	109	110	111	112	113	114	115	116	117
Arg	· His	· Pro	· Gly	· Asn	· Phe	· Gly	· Ala	· Asp	· Ala	· Gln	· Gly	· Ala	· Met	· Asn
118	119	120	121	122	123	124	125	126	127	128	129	130	131	132
Lys	· Ala	· Leu	· Glu	· Leu	· Phe	· Arg	· Lys	· Asp	· Ile	· Ala	· Ala	· Lys	· Tyr	· Lys
133	134	135	136	137	138	139	140	141	142	143	144	145	146	147
Glu	· Leu	· Gly	· Tyr	· Gln	· Gly	Carboxyl end								
148	149	150	151	152	153									

the amino acid sequence of myoglobin from the sperm whale. This protein has a molecular weight of 16,900 and contains 153 amino acid residues in a single polypeptide chain. From research on the amino acid sequence of homologous proteins of different species (i.e., the cytochrome *c* molecules isolated from various plants and animals), it has been found that the amino acid units of polypeptide chains can be divided into two classes. At some positions in the polypeptide chain the amino acid residues are identical in all species. These amino acid residues appear to be required for the characteristic function of the protein molecule. In contrast, at other positions in the polypeptide chain the amino acid residues may vary from one species to another. It has therefore been concluded that certain amino acids in specific positions in the chain are essential for the function of the protein, whereas others are specific for the species of organism. Thus the amino acid sequence of proteins has two kinds of biological meaning.

From these considerations we see that the biosynthesis of different protein molecules must be carried out with great precision by the cell, in order to preserve not only the capacity of the proteins to fulfill their intended functions but also to preserve their species individuality. Moreover, we have also seen (Chapter 7) that protein synthesis may require up to 90 per cent of the cell's biosynthetic energy, at least in bacteria and other fast-growing cells. For these reasons the synthesis of proteins is perhaps the most elaborate and complex biosynthetic process occurring in living organisms.

8-7 THE SYNTHESIS OF PROTEINS

The first stage in protein synthesis is the enzymatic activation of the twenty different amino acids, at the expense of ATP energy, to form derivatives, called amino acyl-transfer RNA's, which have two significant functions. First, they are energy-rich compounds that serve as carriers or donors of the amino acyl group

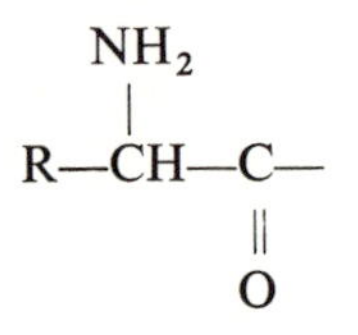

Secondly, these derivatives serve as "adapters" in translation of the 4-letter language of nucleic acids into the 20-letter language of proteins.

For the activation reaction, which takes place in the soluble portion of the cytoplasm, twenty different amino acid activation enzymes are required, each specific for one of the twenty amino acids. In this reaction each amino acid reacts to form a high-energy linkage with its corresponding specific carrier molecule, which consists of a specialized type of RNA called transfer RNA. Transfer RNA's are relatively small RNA molecules, with molecular weights in the range 23,000–30,000; they contain from 75 to 90 nucleotide residues, many of which contain derivatives of the normal purine and pyrimidine bases. The structure of all tRNA molecules may be designated by the "cloverleaf" model in Fig. 8–10. Each tRNA has a terminal adenylic acid residue that serves as the point of attachment of the amino acid and another specific site made up of a nucleotide triplet, the *anticodon*, which is different for each tRNA. There is at least one specific tRNA molecule for each amino acid, but most amino acids have more than one type of transfer RNA to which they can be enzymatically linked. In fact, the amino acid leucine has five different tRNA's in the yeast cell. Specific types of transfer RNA molecules are symbolized by subscripts showing the amino acid for which they are specific, for example, $tRNA_{Ala}$, $tRNA_{Lys}$, or $tRNA_{Glu}$. Thus, for each amino acid there is a specific activating enzyme and at least one specific

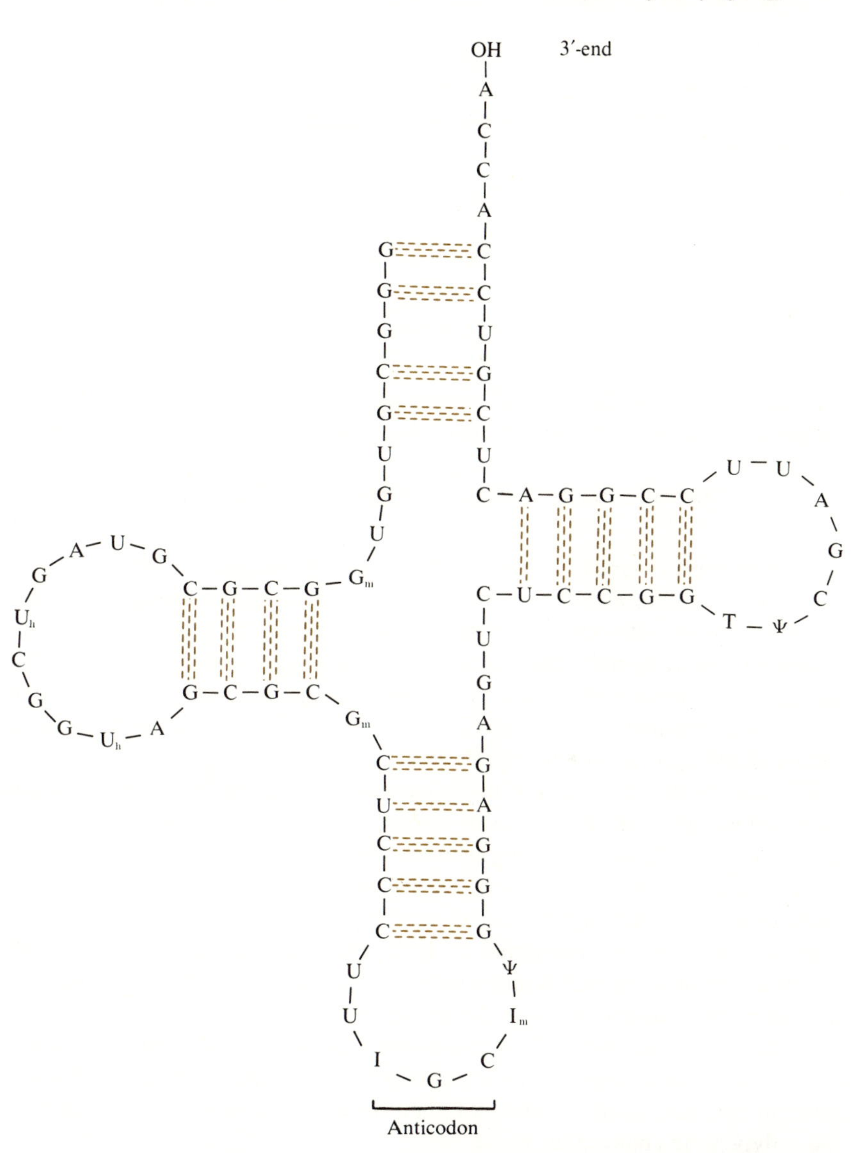

Figure 8–10. Alanine tRNA of yeast. Through base-pairing within the chain, the tRNA molecules can assume the cloverleaf structure shown above. The symbols ψ, Im, I, Gm, and U_h designate minor bases that are characteristic of tRNA molecules.

tRNA to which the amino acid becomes enzymatically attached. The activating reaction for the amino acid alanine proceeds as follows

$$\underset{\text{alanine}}{CH_3\text{—}\overset{\overset{NH_2}{|}}{C}H\text{—}COOH} + ATP + tRNA_{Ala} \xrightarrow[\text{enzyme Ala}]{\text{activating}}$$

$$\underset{\text{alanyl-tRNA}_{Ala}}{CH_3\text{—}\overset{\overset{NH_2}{|}}{C}H\text{—}\underset{\underset{O}{\|}}{C}\text{—}tRNA_{Ala}} + AMP + \text{pyrophosphate}$$

The activation reaction, which causes pyrophosphate cleavage of ATP, is followed by the hydrolysis of the pyrophosphate to yield inorganic phosphate

$$\text{pyrophosphate} + H_2O \xrightarrow{\text{pyrophosphate}} 2\ \text{phosphate}$$

Two high-energy bonds of ATP are thus used up to create the covalent union between the amino acid and its tRNA. The new chemical linkage formed is an ester linkage between the carboxyl group of the amino acid and a hydroxyl group of the terminal adenylic acid residue of the tRNA molecule. The standard free energy of hydrolysis of this linkage is about −6.5 kcal; it is therefore a high-energy linkage. The structure of the terminal portion of alanyl-$tRNA_{Ala}$ is shown in Fig. 8–11.

In a similar manner, each of the twenty different amino acids is enzymatically linked by an ester bond to its corresponding tRNA molecule; for each amino acid a different activating enzyme is required. Transfer RNA molecules are thus carriers of amino acids, just as uridine nucleotides are carriers of sugars and coenzyme A is carrier of acyl groups. The activated amino acids are now ready to be assembled to form the polypeptide chain; we shall presently see how the tRNA molecule acts as an adapter to make possible the arrangement of amino acids in the proper sequence.

In the next stage of protein synthesis the messenger RNA from the nucleus and the activated amino acids attached to their corresponding tRNA's are brought together on the surface of the ribosomes, where the formation of the polypeptide chain takes place.

Ribosomes are granular elements in the cytoplasm of all cells; they are easily isolated by differential centrifugation of cell extracts. In prokaryotic cells, such as *E. coli*, the ribosomes have a particle weight of about 2.8 million and a diameter of about 180 Å and they are found in the free state in the cytoplasm. In eukaryotic cells the ribosomes are somewhat larger, with a particle

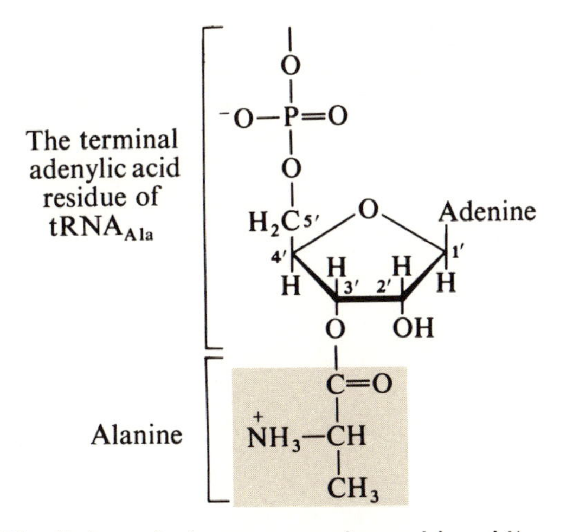

Figure 8–11. The linkage between an amino acid and its corresponding tRNA.

weight of about 4.2 million, and a large fraction of them is bound to the membrane of the endoplasmic reticulum. The ribosomes of both prokaryotic and eukaryotic cells are otherwise very similar in composition and function.

E. coli ribosomes contain about 35 per cent protein and 65 per cent ribonucleic acid, which occurs in three special types known as 23 S, 16 S, and 5 S ribosomal RNA, on the basis of their rates of sedimentation in a centrifugal field; the symbol S stands for the svedberg, the unit in which sedimentation rates are expressed. Ribosomes have a large and a small subunit. The small subunit (molecular weight, 1,000,000) contains one molecule of 16 S rRNA and twenty different polypeptide chains. The large subunit (molecular weight, 1,800,000) contains one molecule of 23 S and one of 5 S rRNA plus about thirty polypeptide chains. The major and minor subunits of ribosomes are associated during the synthesis of a polypeptide chain, but they come apart after the chain has been completed. Fig. 8–12 summarizes the structural features of ribosomes.

The biosynthesis of a polypeptide chain begins by the binding of the mRNA molecule specifying a given protein to the small ribosomal subunit, in such a way that the first or initiating codon is positioned on a specific site on the ribosomal subunit, called the peptidyl or P site. Then the initial amino acyl-tRNA, whose identity we shall discuss later, also binds to the ribosome, positioned next to the first codon of the messenger RNA. The amino acyl-tRNA is recognized and selected because it has an anticodon triplet that is complementary to the codon in the bound mRNA molecule, and binds to it through hydrogen-bonding of their complementary base pairs. In this selection and binding process the amino acid attached to the end of the tRNA

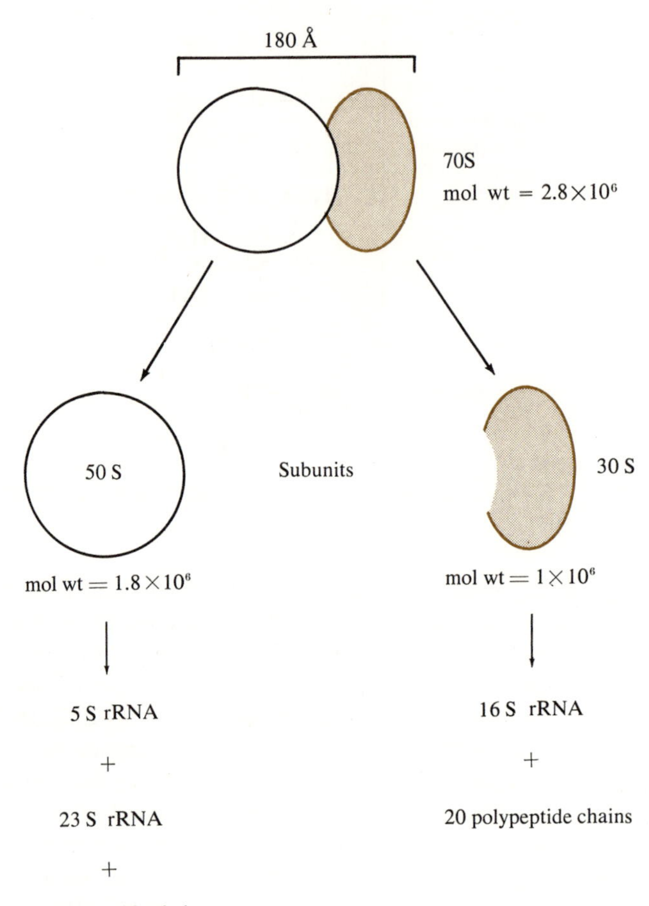

Figure 8–12. Structure of *E. coli* ribosomes.

molecule plays no role; the specificity of the amino acyl-tRNA lies entirely in its anticodon. After the initiating amino acyl-tRNA has been positioned to the codon in the bound mRNA, the large ribosomal subunit combines with the smaller; the resulting structure is the "initiation complex". It is now ready to bring about the successive elongation steps by which the polypeptide chain is built. In the elongation process, the incoming amino acid units are always attached to the carboxyl group of the preceding amino acid residue; thus the synthesis of the polypeptide chain starts with the amino terminal end and proceeds to the carboxyl-terminal end (Fig. 8–13).

The sequential elongation of the polypeptide chain proceeds in three major steps for each amino acid residue added. In the first, the incoming amino acyl-tRNA specified by the next codon of the mRNA is selected and bound to the amino acyl or A site, positioned by hydrogen bonding of its anticodon to the codon. For this process one molecule of guanosine triphosphate (GTP) must bind to the ribosome. In the second step, the peptide bond is formed between the initiating amino acid at the P site and the second amino acid at the A site, by the action of the enzyme *peptidyl transferase*, which causes the transfer of the initiating amino acyl group from its tRNA on the P site to the free amino group of the new amino acyl-tRNA on the A site. The result is a dipeptide attached by its carboxyl group to the tRNA of the second amino acid. The now empty tRNA of the initiating amino acid remains attached to the P site, but only temporarily, as we shall see.

In the third step of the elongation process, the dipeptidyl-tRNA is translocated from the A site to the P site, in order to "clear" the A site for the next incoming amino acyl-tRNA; simultaneously, the mRNA is moved along the ribosome by one codon. This translocation process requires a molecule of GTP, which is split to GDP and phosphate. As a result of the translocation of the peptidyl-tRNA from the A site to the P site, the empty tRNA left behind on the P site is "bumped" off the ribosome. The complex of peptidyl-tRNA, mRNA, and the ribosome is now ready to receive the third amino acyl-tRNA. This cycle of three steps must be repeated for each amino acid unit added to the polypeptide. As each amino acid unit is inserted, the mRNA molecule and the growing polypeptide chain, attached by its terminal carboxyl group to the corresponding tRNA, must be moved along the ribosome by one codon. The mRNA and the growing polypeptide chain "track" through the groove between the major and minor subunits of the ribosome. Since most polypeptide chains are between 100 to 300 residues long, the mRNA and growing polypeptide chain must be pushed along the ribosome many times; with each translocation a molecule of GTP undergoes hydrolysis to GDP and phosphate.

At the end, when the polypeptide chain approaches completion, special termination codons (see below) in the mRNA signal the end of the chain. When this signal is received, a special enzyme comes into play that now releases the complete polypeptide chain in free form. Simultaneously, the two subunits of the ribosome come apart in free form, ready to start a new chain.

In the intact cell a single messenger RNA molecule may be used as a template by several or many ribosomes simultaneously, each of them making its own polypeptide chain independently of the others. For example, the mRNA molecule coding for the synthesis of the polypeptide chains of hemoglobin in the immature, growing red blood cell, can be utilized by at least five ribosomes simultaneously. Such a cluster of ribosomes, all engaged in translating from the same mRNA molecule, is called a polyribosome or polysome.

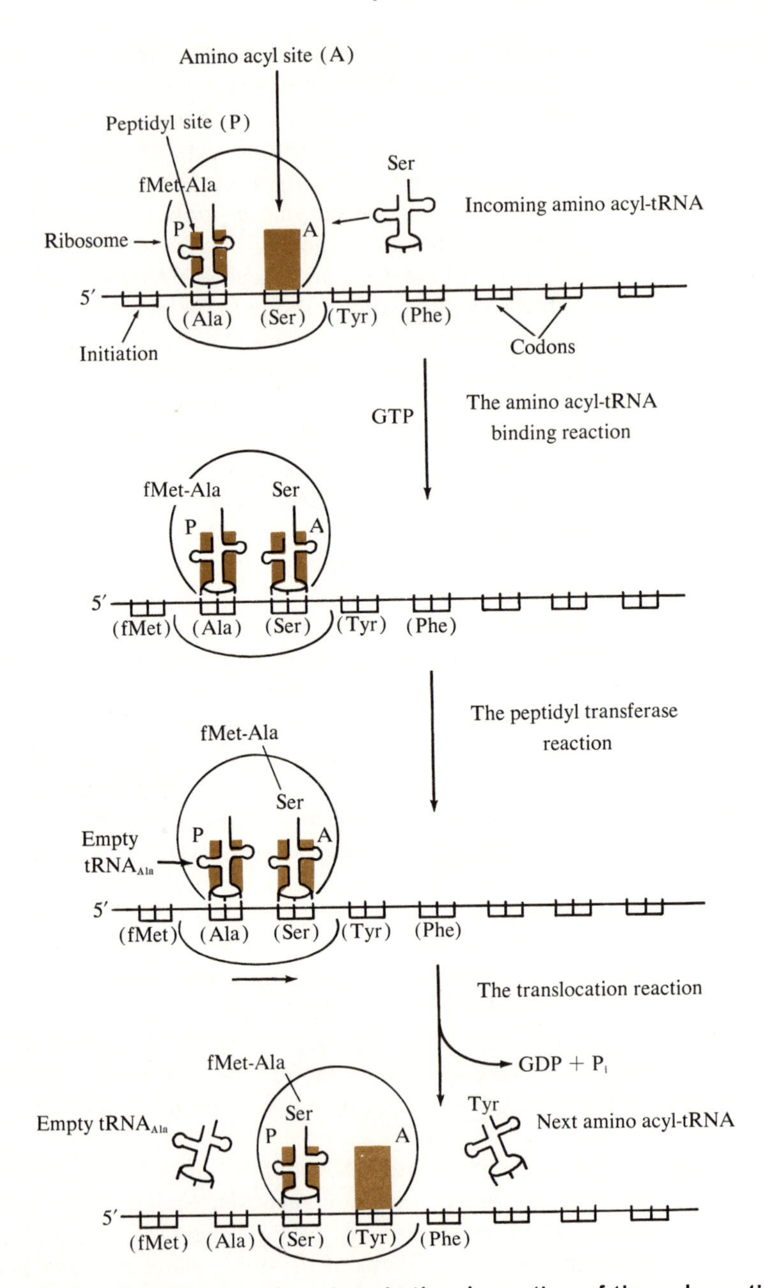

Figure 8–13. The three major steps in the elongation of the polypeptide chain (from Albert L. Lehninger, *Biochemistry*, Worth Publishers, New York, 1970, p. 701).

Now let us examine the energy balance sheet for protein synthesis. In order to make a polypeptide chain of 100 residues, we need to activate 100 amino acid molecules by attaching them to their corresponding tRNA molecules. For this we need 100 ATP molecules, but this process will utilize a total of 200 high-energy phosphate bonds. For each amino acid residue inserted, we must move the polypeptide chain and the mRNA along the ribosome by one codon; for each translocation step one molecule of GTP is hydrolyzed to GDP and phosphate. Thus we need a total of 300 high-energy phosphate bonds to make a polypeptide chain from 100 amino acid molecules. Since each peptide bond requires a minimum input of about 5.5 kcal, and three high-energy phosphate bonds of ATP (GTP) can yield $3 \times 7.3 = 21.9$ kcal, the synthesis of a peptide bond would appear to waste much free energy. But this is not really wasted energy; it is the price the cell must pay to ensure that protein synthesis goes overwhelmingly to completion, thus guaranteeing that the translation of genetic information into protein structure proceeds with very high accuracy.

8-8 THE GENETIC CODE

Now let us see how the four different bases of DNA, which constitute its "language," can code for the twenty different amino acids, which constitute the language of protein structure. It had long been appreciated that a group of at least three bases must be required to code each amino acid, since four bases arranged in groups of two can yield only $4^2 = 16$ different combinations, which is insufficient to code twenty amino acids, whereas arranged in groups of three they can yield $4^3 = 64$ different combinations, or more than enough for the twenty amino acids. However, it was not known whether "commas" are used to separate the coding groups of nucleotides in DNA as punctuation. However, in the late 1950's it was proven by genetic experiments that each amino acid is specified by a triplet of nucleotides and that the genetic code is commaless. Thus the codons for the different amino acids must be read in proper register, that is, in a specific "reading-frame" relationship.

The identification of the code-word triplets specifying each amino acid was made possible by two discoveries of Nirenberg and his colleagues. In the first, Nirenberg and Matthaei found that isolated *E. coli* ribosomes will utilize as messenger RNA not only naturally occurring mRNA but also synthetically prepared polyribonucleotides of known base composition and sequence. For example, they found that a polyribonucleotide consisting only of uridylic acid residues, called polyuridylic acid or poly U, when used as an artificial messenger with *E. coli* ribosomes codes for a polypeptide chain containing only phenylalanine residues. Similarly, polyadenylic acid codes for a polypeptide chain containing only lysine residues. This discovery made it possible to determine the bases present in each codon, but did not allow firm conclusions as to the sequence of bases in each codon. However, in a second major discovery Nirenberg and Leder found that *E. coli* ribosomes in the absence of an energy source will bind simple trinucleotides. When they do so, they also will bind the corresponding amino acyl-tRNA. For example, when the trinucleotide designated as UUU is bound by the ribosome, it will also bind phenylalanyl-$tRNA_{Phe}$, but no other amino acyl-tRNA. By using synthetically prepared trinucleotides of known base composition and sequence, Nirenberg, and also Khorana, were able to establish the identity of all the triplet code words for the amino acids within a remarkably short period of time, in one of the greatest experimental advances in modern

TABLE 8–2. The dictionary of genetic code words for the amino acids. The codons read in the 5′ ⟶ 3′ direction. The nonsense codons, which are now known to be termination signals, are printed in color.

	U	C	A	G
U	UUU Phe UUC Phe UUA Leu UUG Leu	UCU Ser UCC Ser UCA Ser UCG Ser	UAU Tyr UAC Tyr UAA Ochre UAG Amber	UGU Cys UGC Cys UGA UGG Trp
C	CUU Leu CUC Leu CUA Leu CUG Leu	CCU Pro CCC Pro CCA Pro CCG Pro	CAU His CAC His CAA Gln CAG Gln	CGU Arg CGC Arg CGA Arg CGG Arg
A	AUU Ile AUC Ile AUA Ile AUG Met	ACU Thr ACC Thr ACA Thr ACG Thr	AAU Asn AAC Asn AAA Lys AAG Lys	AGU Ser AGC Ser AGA Arg AGG Arg
G	GUU Val GUC Val GUA Val GUG Val	GCU Ala GCC Ala GCA Ala GCG Ala	GAU Asp GAC Asp GAA Glu GAG Glu	GGU Gly GGC Gly GGA Gly GGG Gly

scientific history. The complete dictionary of the code words is shown in Table 8–2.

We will note that most of the amino acids have more than one code word, but no code word specifies more than one amino acid. When there are multiple code words usually the first two letters are the same in all. For example, the code words for alanine are GCU, GCC, GCA, and GCG; the first two nucleotides GC are common to all. This has led to the view that the first two letters provide most of the specificity of the triplet. Moreover, it has been found that this code-word dictionary is identical in the chromosomes of man, *E. coli*, and certain amphibia, plants, and viruses; it thus appears to be universal among all species.

Two major questions remain to be answered. How does the ribosome recognize the initiating codon that begins the code for a polypeptide chain? How does the ribosome know when to stop the construction of a polypeptide chain? Clearly, if there were no signal in the mRNA molecule indicating when the code for a given polypeptide chain begins, the ribosome might begin translation at random points along the mRNA, and thus yield defective or incomplete proteins. Conversely, if there are no signals to terminate a chain, it would be elongated indefinitely, until the end of the mRNA is reached. Recall that an mRNA molecule may code for several polypeptide chains.

The signals for initiation of polypeptide chains are not yet entirely understood, but it has been firmly established that the synthesis of all polypeptide chains in prokaryotic cells begins with a derivative of the amino acid methionine, namely, *N*-formylmethionine

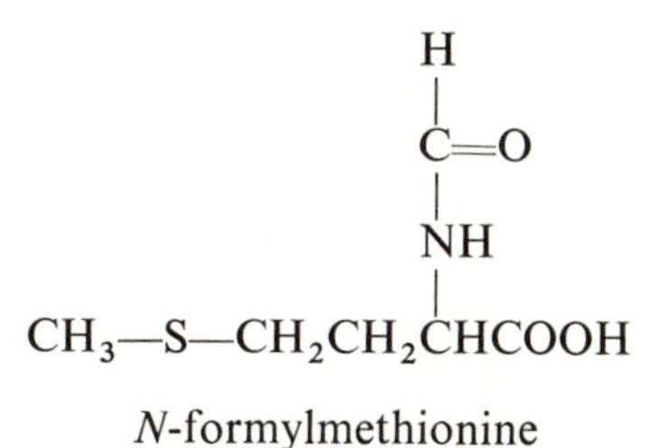

N-formylmethionine

In eukaryotic cells, polypeptides appear to begin with the amino acid methionine. However, it is not entirely clear how these initiating amino acids are signaled by the mRNA molecule.

More is known about the signals for termination of polypeptide chains. After the genetic code words for the different amino acids were identified, it was found that there are three triplets that do not code for any amino acid. These are UAA, UAG, and UGA. The first two are called ochre and amber, respectively. For a long time these three triplets were called nonsense codons, but they are now known to be signals for the termination of polypeptide chains. Whenever they occur, the elongation of the polypeptide chain comes to a stop and the chain is released from the ribosome.

9

THE ASSEMBLY OF CELL STRUCTURE

We have now seen how the major chemical components of cells are constructed from simple precursor molecules at the expense of phosphate bond energy. Now we come to the final step in biosynthesis: the assembly of proteins, nucleic acids, lipids, and other biomolecules into supramolecular systems, such as membranes and enzyme complexes, then into organelles such as mitochondria and endoplasmic reticulum, and these in turn into living, differentiated cells, each with its own distinctive genetic makeup.

In this chapter we will first consider the relationship between information and entropy, which will tell us that the assembly of highly ordered, three-dimensional cell structure from a collection of randomly disposed precursor molecules is a process that must proceed with a very large decrease in internal entropy. Then we shall consider the major question posed in this chapter: How is the one-dimensional or linear information in the base sequence of DNA translated into the three-dimensional structure of the living cell?

9-1 INFORMATION AND ITS MEASUREMENT

A whole new field of science has grown around the concept of information and its measurement, storage, and communication. This new field of information theory is an extension of thermodynamics and of probability theory. Its concepts have been of practical use in the development of computers and communication networks, and they are being applied with profit to analysis of the structure and function of living organisms. In the formalism of information theory, the basic quantitative unit of information is the *binary digit*, abbreviated as the *bit*. The amount of information in a message is expressed in terms of the probability of a sequence of binary choices or bits. For example, if someone has to choose one specific card from a row of sixteen in order to win a prize, the probability (Po) of choosing the right one is only one out of sixteen or 1/16. But suppose the chooser makes the correct choice on his first trial because he received the necessary information from someone who knew the location of the card. After receiving this information the probability (P) of choosing the correct card is 1.0. The ratio of the probabilities (P/Po) is thus 16. We may now calculate the amount of information in the message received by the chooser. It is given by the expression

$$I = \log_2 \frac{P}{Po}$$

where I is *information* in *bits* and $\log_2$ is the logarithm to the base 2. Now $2^4 = 16$, so the logarithm to the base 2 of P/Po ($= 16$) is 4. Therefore $I = 4$, and the message is said to contain four bits of information. This means that four correct binary choices would have to be made to choose the correct card. If the correct choice of one card is to be made from thirty-two cards, then $\log_2 32$ or five bits of information are needed. Since simple numerical information of this kind can be expressed as a consecutive sequence of binary yes–no choices, it can easily be coded or stored in linear form, as on a tape.

9-2 INFORMATION CONTENT OF CELLS

Many attempts have been made to calculate the amount of information in the linear base sequence of DNA or the amino acid sequence of a protein by calculations such as those described above. However, this approach has not been successful since the information inherent in the sequence of DNA or proteins cannot as yet be reduced to simple consecutive binary units. For one thing, the twenty different amino acid units of proteins probably have more than one "meaning" in protein structure. Moreover, the information in any given amino acid unit may have different levels of meaning, which are significant only as part of a much longer sequence of amino acids. Nevertheless, even with many oversimplifications it has been calculated that a single *E. coli* cell contains an enormous amount of information; one "guesstimate" is that it may greatly exceed 10^{12} bits.

But we can make the point that an *E. coli* cell contains a very large amount of information in a more graphic and understandable manner. In Fig. 9–1 is

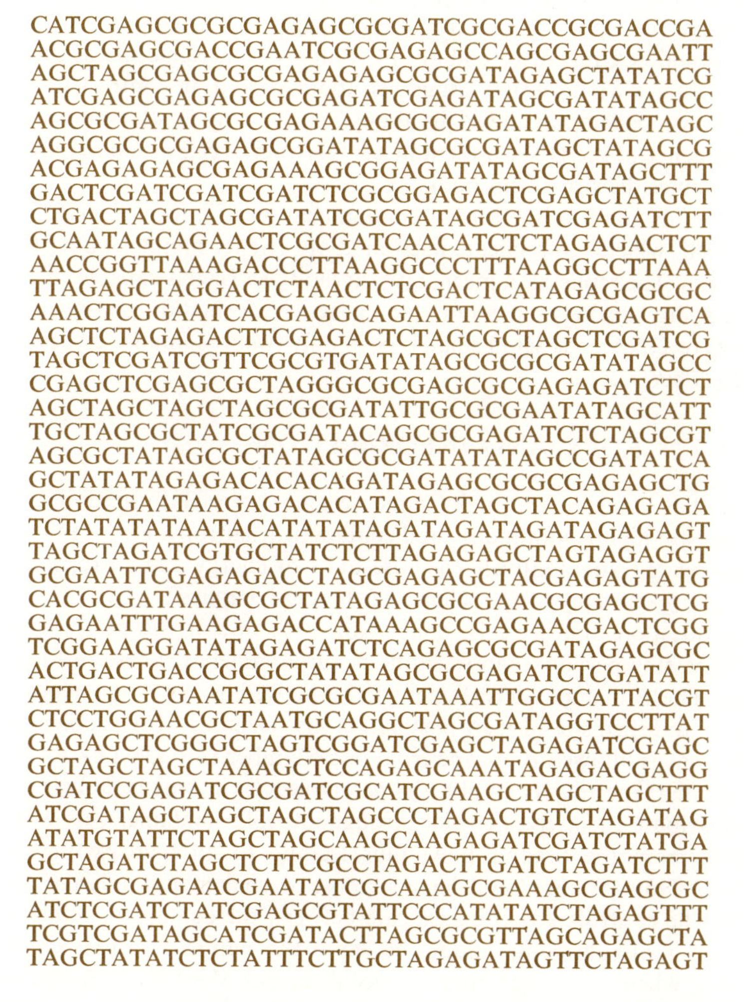

Figure 9–1. An imaginary base sequence for a segment of DNA having a molecular weight of about 1,000,000.

shown an imaginary base sequence for a segment of DNA having a molecular weight of about 1,000,000, containing about 1500 base pairs. Now let us suppose we were to use the same size type to write out the base sequence of the entire chromosome of an *E. coli* cell, which consists of a single DNA molecule of molecular weight 2,800,000,000, corresponding to about 4,700,000 base pairs. We would require about 3000 pages of this closely printed type, which would be equivalent to about 14 volumes of the same size as this book. These volumes would weigh altogether about 9000 g, in contrast to a single *E. coli* chromosome, which weighs about 1×10^{-14} g. On the same basis, the base sequence of the DNA present in a single human cell would require about 7000 volumes the size of this book.

Now let us consider the rate of communication of information by a single *E. coli* cell, which can replicate itself in about twenty minutes under the best conditions. The replication mechanism in the *E. coli* cell is capable of reading and copying precisely an amount of genetic information contained in 14 volumes of DNA code set in small type in only twenty minutes, or better than 2.5 pages per second.

9–3 ENTROPY AND INFORMATION

To say that living cells contain immense amounts of information is to say that they are highly nonrandom and contain little entropy. How can information be related quantitatively to those units of energy and entropy that are the working currency of the science of thermodynamics?

This question first arose in connection with one of the most celebrated and interesting paradoxes in scientific history, namely the case of Maxwell's demon. The British physicist James Clerk Maxwell once posed a hypothetical problem and then solved it by postulating his "demon." The problem is this. When a gas is allowed to escape from one chamber through a narrow orifice into another one that is empty, the gas molecules normally diffuse so that they distribute themselves evenly throughout both containers, as would be expected from the Second Law, thus leading to an increase in entropy (Fig. 9–2). Now, Maxwell said, let us suppose that a demon stands at the orifice and that he controls a frictionless gate between the containers. When a fast or "hot" molecule comes along, he lets it through to the other container. However, when a slow or "cold" molecule comes, the demon closes the gate and does not allow it to enter. In this way, only hot molecules can go into the empty chamber and the cold molecules must stay behind. Such a process would *decrease* the entropy or randomness of the system without changing its total energy (Fig. 9–2); it would thus flout the Second Law. Maxwell pointed out that since the demon does no work but merely selects the hot molecules such a process could theoretically occur, even though there are no recorded instances of it happening.

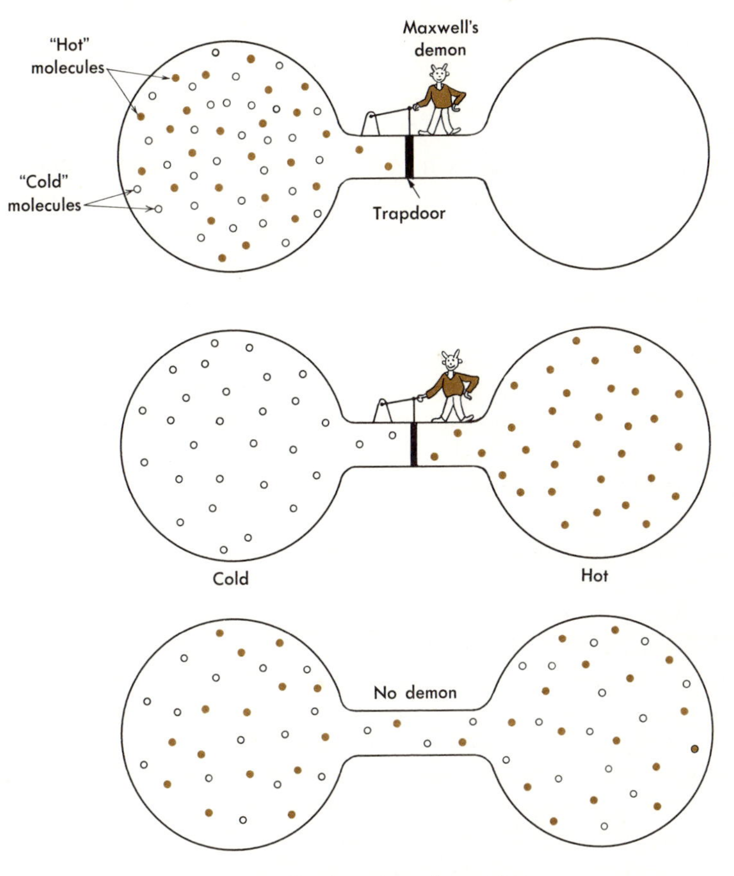

Figure 9–2. Maxwell's demon can separate hot from cold molecules only if he has information.

For years there seemed to be no reasonable explanation for this paradox. However, Szilard pointed out in 1929 that Maxwell's demon requires the use of information in order to select the hot molecules. He postulated that the information required to select the molecules is energetically equivalent to the decrease of entropy that occurs when only the hot molecules are allowed to enter the second chamber. From this key idea an inverse numerical correspondence between information and entropy has been deduced: about 10^{23} bits of information are required to reduce the entropy of a system by 1.0 kcal/mole degree. It is not yet possible to describe a similar quantitative relationship between genetic coding units and entropy units, but whatever the numerical correspondence, it is clear that cells contain relatively little entropy or randomness. Moreover, whenever a cell is formed from simple precursors in the environment, there must be a large local decrease in entropy.

9-4 THE PROGRAMMING OF CELL STRUCTURE

Because the spontaneous growth of a cell from simple building-block molecules is such a highly improbable event from the thermodynamic and statistical point of view, there must be some underlying simplicity in the replication and growth of cells, some feature that makes the self replication of cells highly probable and even inevitable, providing that the parent cell provides a specific program for the development of the cell. Such a program would be necessary to assure that all the enzyme-catalyzed reactions and physical events required to construct a cell proceed in one unique sequence, in such a way that each reaction step leads inevitably to the next and that there is but one possible end product of this sequence, namely, an intact, living cell.

From these considerations it is evident DNA must also serve as the programming system for orderly replication of the entire cell. It must specify not only which protein molecules are made but also how many of each and in what sequence. The DNA must also specify the three-dimensional configuration of each protein as well as its specific biological activity. Furthermore, the DNA must also program the assembly of specific protein molecules into organized clusters or supramolecular complexes, such as multienzyme complexes, membranes, and ribosomes. Ultimately the DNA must also play a role in specifying the formation of cell organelles, and in directing their assembly to form a complete cell. To go further, the DNA must also specify whether a cell is to become a nerve cell in one case or a kidney cell in another.

We do not yet know the details of all these DNA-directed events for any type of cell. However, two basic first principles underlying the development of cell structure have emerged. They are implicit in the answers to the following questions: How is the linear or one-dimensional information of DNA converted into the three-dimensional information inherent in cell structure? In what way is energy transformed to create three-dimensional order from a one-dimensional message? We shall approach these questions by returning to proteins and the determination of their structure.

9-5 THE THREE-DIMENSIONAL STRUCTURE OF PROTEINS

In Chapter 8 we left the newly synthesized polypeptide chain, with its genetically directed sequence of amino acids, as it peeled off the ribosome. In this form it is not yet a finished protein since we know today that each protein molecule not only has a distinctive amino acid sequence, but also a distinctive three-dimensional configuration into which its polypeptide chain is coiled and folded. It is now certain that this specific folded arrangement of a native protein molecule is necessary for its biological specificity or activity. For example, the catalytic activity of enzymes depends on a specific coiled or folded configuration of the polypeptide chain(s) in the enzyme molecule. An enzyme molecule may be caused to unfold into a randomly coiled polypeptide chain, a process that is called denaturation, by simply heating it to 100°C or by treating it with excess acid or base. As a result of this unfolding process, the catalytic activity of the enzyme molecule will be lost, yet the covalent backbone of the polypeptide chain remains intact and its amino acid sequence is preserved. The biological activity of protein molecules thus requires a specific kind of three-dimensional arrangement of their polypeptide chains.

Proteins fall into two major classes, *fibrous* and *globular*, depending on their three-dimensional structure. The fibrous proteins are water-insoluble proteins serving a structural role; they are the main components of such fixed structures as hair, nails, feathers, skin, and tendons. Among the fibrous proteins are the α-keratin of hair, the β-keratin of silk, and the collagen of connective tissue. All the fibrous proteins have their polypeptide chains coiled or folded in a regular manner along one dimension. One such arrangement is the α-helix, characteristic of the keratin in hair; another is the β-configuration, a zig-zag arrangement of the polypeptide chain found in fibroin, the protein component of silk.

The globular proteins, on the other hand, are more nearly spherical in shape, are usually soluble in water, and they diffuse readily; they play a dynamic rather than a structural role. Practically all the known enzymes are globular proteins. In globular proteins the long polypeptide chain(s) are tightly folded in such a way as to yield a spherical or globular conformation. However, it is one of the important conclusions of modern research on the structure of globular protein molecules by x-ray crystallography that the polypeptide chain is not simply folded in a random manner and rolled up into an amorphous ball; on the contrary, the folding of the chain is exact and precise, to yield a rigid, stable structure. Each protein molecule of a given kind is exactly like the next in its steric conformation.

Figure 9–3 shows the three-dimensional structure of the globular protein myoglobin, deduced from its x-ray diffraction pattern by Kendrew, a British molecular biologist. Note the complexity with which the backbone and side chains of the polypeptide chain are folded and convoluted to yield a globular conformation of the molecule, within which there is very little open space. Although this arrangement may appear irregular and amorphous, it is entirely specific; each native myoglobin molecule is folded in precisely the same way as the next. Moreover, we know today, from the results of x-ray analysis of many other proteins, such as cytochrome *c*, chymotrypsin, and ribonuclease, that each type of globular protein has a distinctive conformation

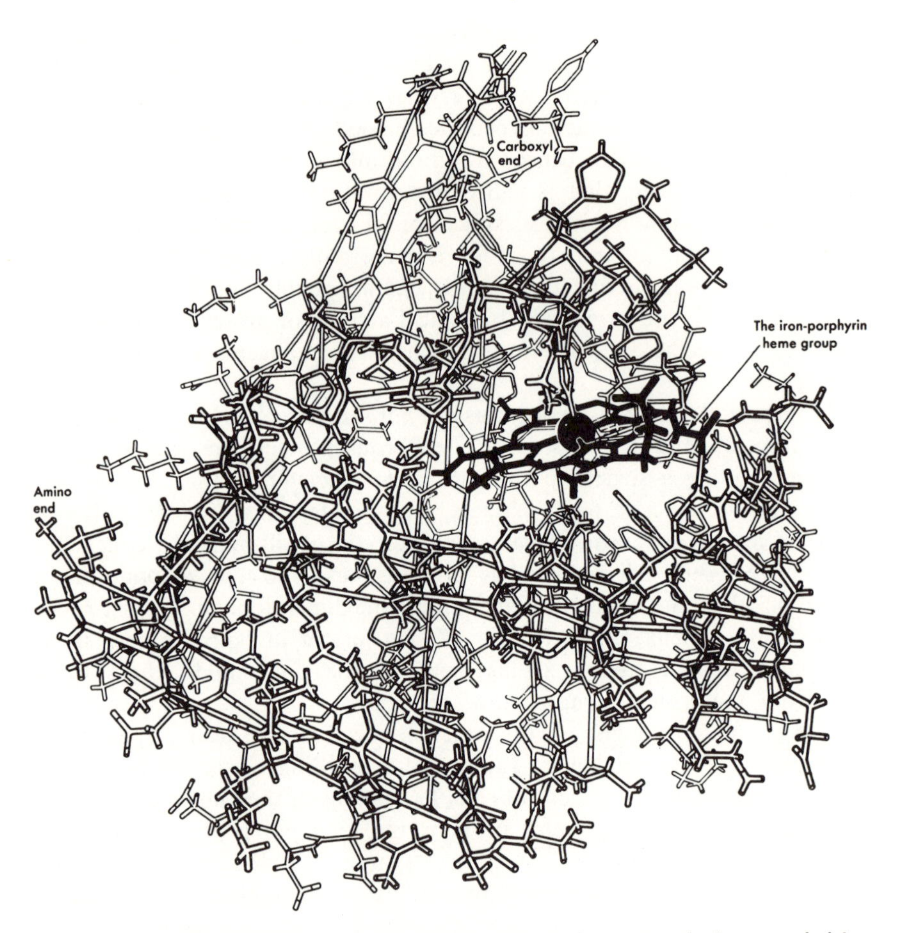

Figure 9–3. The three-dimensional structure of sperm whale myoglobin (from "The Three-dimensional Structure of a Protein Molecule" by John C. Kendrew. Copyright © December 1961 by Scientific American, Inc. All rights reserved).

in which its polypeptide chain(s) is folded. The characteristic biological activity of each of these proteins is determined by the conformation of its polypeptide chain.

9–6 THE FORMATION OF THREE-DIMENSIONAL STRUCTURE

Recent research indicates that the specific three-dimensional structure of globular proteins, that is, the specific manner in which the polypeptide chain is folded, is the automatic consequence of their amino acid sequence. In

Table 8–1 we saw the amino acid sequence of myoglobin. This sequence, which is linear or one dimensional, contains the message for the characteristic three-dimensional folded conformation of myoglobin shown in Fig. 9–3. But how can a one-dimensional sequence of amino acids specify a three-dimensional conformation?

The answer is to be found in the R groups or side chains of the twenty different amino acid residues. As we have seen (Chapter 8), these side chains differ in size, shape, polarity, and electrical charge. They are spaced so closely together in the polypeptide chain that they tend to interact with each other, so that any given side chain either attracts or repels its nearest neighbor side chain. For example, if a lysine side chain, which has a positive charge at pH 7.0 (Fig. 8–8) is next to a glutamic acid residue, which has a negative charge, they will attract each other. But two adjacent lysine residues will repel each other. Such interactions between neighboring side chains will subject the backbone of the polypeptide chain to mechanical stress and force it into a configuration in which this stress will be relieved. Similarly, if there are adjacent valine and isoleucine residues, which are especially bulky, the backbone of the chain will also be bent into a compensating configuration. Especially important are those side chain interactions in which hydrogen bonds form between one side chain and another several residues removed. This type of interaction causes the polypeptide backbone to coil into the α-helical arrangement (Fig. 9–4), with the successive loops held together by hydrogen bonds. The α-helix is characteristic of the α-keratin class of fibrous proteins.

As a result of such interactions among the R groups and between the R groups and the surrounding water, the backbone of the polypeptide chain becomes coiled or folded. In this manner, residue by residue, the entire polypeptide chain folds into a characteristic three-dimensional arrangement.

No external work or enzyme action must be exerted on a polypeptide chain to fold it into its characteristic conformation. It folds itself quite spontaneously and automatically, because the successive side chains of the amino acid units and the backbone tend to seek the most stable possible configuration, that with the least free energy. This is another way of saying that the polypeptide chain simply is seeking its equilibrium state, at which its free energy is minimum.

It may seem paradoxical at first that a polypeptide chain should automatically and spontaneously assume a complex, specific configuration in which it has biological activity, since in this process the chain has undergone a transition from a relatively random, disordered, stringy molecule, having much entropy, into a more ordered, more specific, and thus less random configuration, one evidently possessing relatively little entropy. Usually we think of physical or chemical processes occurring in such a direction that entropy increases. But the spontaneous folding of the polypeptide chain does *not* represent a violation of the Second Law of thermodynamics. We must

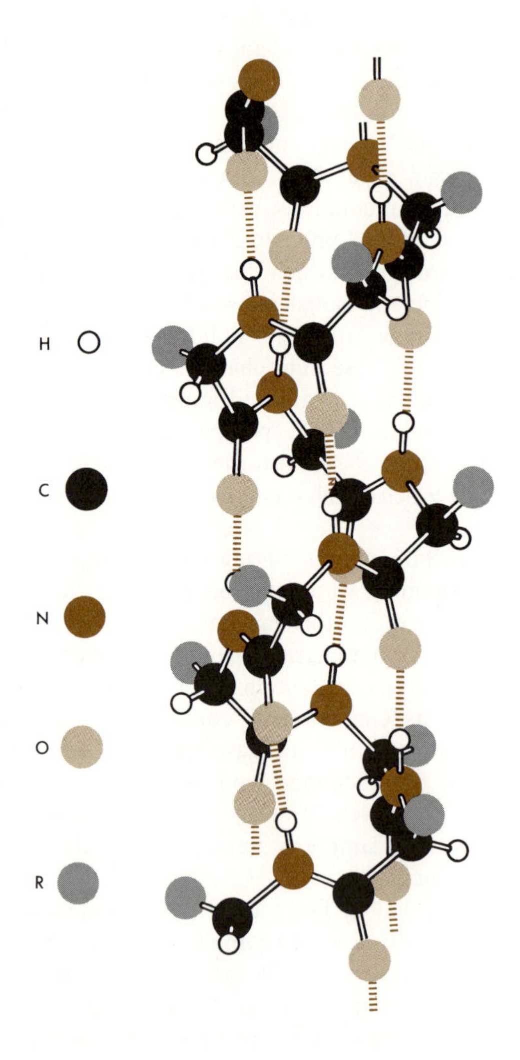

Figure 9–4. The α-helix (redrawn from Linus Pauling, *The Nature of the Chemical Bond*, 3rd ed., Cornell Univ. Press, Ithaca, N.Y., 1960, p. 500).

recall that the Second Law says that it is the entropy of the *universe* that must increase when a process occurs. Although the entropy of the polypeptide chain (which we may regard in the language of thermodynamics as the *system*) decreases as it folds, the surroundings of the polypeptide chain increase in entropy to such an extent that the total entropy in the system plus surroundings (i.e., the universe) increases. It is therefore clear that as a polypeptide chain folds, some other process must occur that leads to an increase in entropy in the surroundings. This external process involves the water molecules in the aqueous medium surrounding the polypeptide chain. When a polypeptide chain is in its random form, it tends to immobilize adjacent water molecules at its hydrophobic or water-hating side chains; these water molecules are held in an ordered, low-entropy configuration. But this is an unstable condition. Because the polypeptide chain has some flexibility, it will spontaneously fold so as to allow the water molecules surrounding the hydrophobic side chains to escape and attain a disordered or high entropy configuration. The folding takes place in such a way that the hydrophobic side chains all become hidden inside the folded structure; only the hydrophilic or water-compatible side chains remain exposed on the surface. Once the chain is folded in this manner, the adjacent water molecules are free to assume a more random state with an increase in entropy. The major driving force for the spontaneous formation of the native folded configuration of a polypeptide chain is therefore the tendency of the surrounding water molecules to seek that configuration in which they have maximum entropy, which occurs when they are maximally hydrogen-bonded to each other. To put it in another way, water molecules like each other more than they like the hydrophobic side chains of polypeptides. Thus the large increase in entropy of the surrounding water more than compensates for the decrease in entropy as the polypeptide chain undergoes folding.

To sum up, the one-dimensional information of DNA is transcribed into the one-dimensional information of messenger RNA, which is in turn translated into the one-dimensional information of the amino acid sequence in the polypeptide chain. But polypeptide chains, unlike DNA and mRNA, are not stable in one-dimensional form, they automatically fold into an information-rich three-dimensional conformation.

9–7 HIGHER-ORDER THREE-DIMENSIONAL STRUCTURES

With the help of the schematic representations in Fig. 9–5 let us now see how the amino acid sequence of polypeptide chains can also direct the formation of higher-order three-dimensional structures. We have already pointed out that many proteins contain more than one polypeptide chain; actually, this is true of most globular proteins. For example, hemoglobin

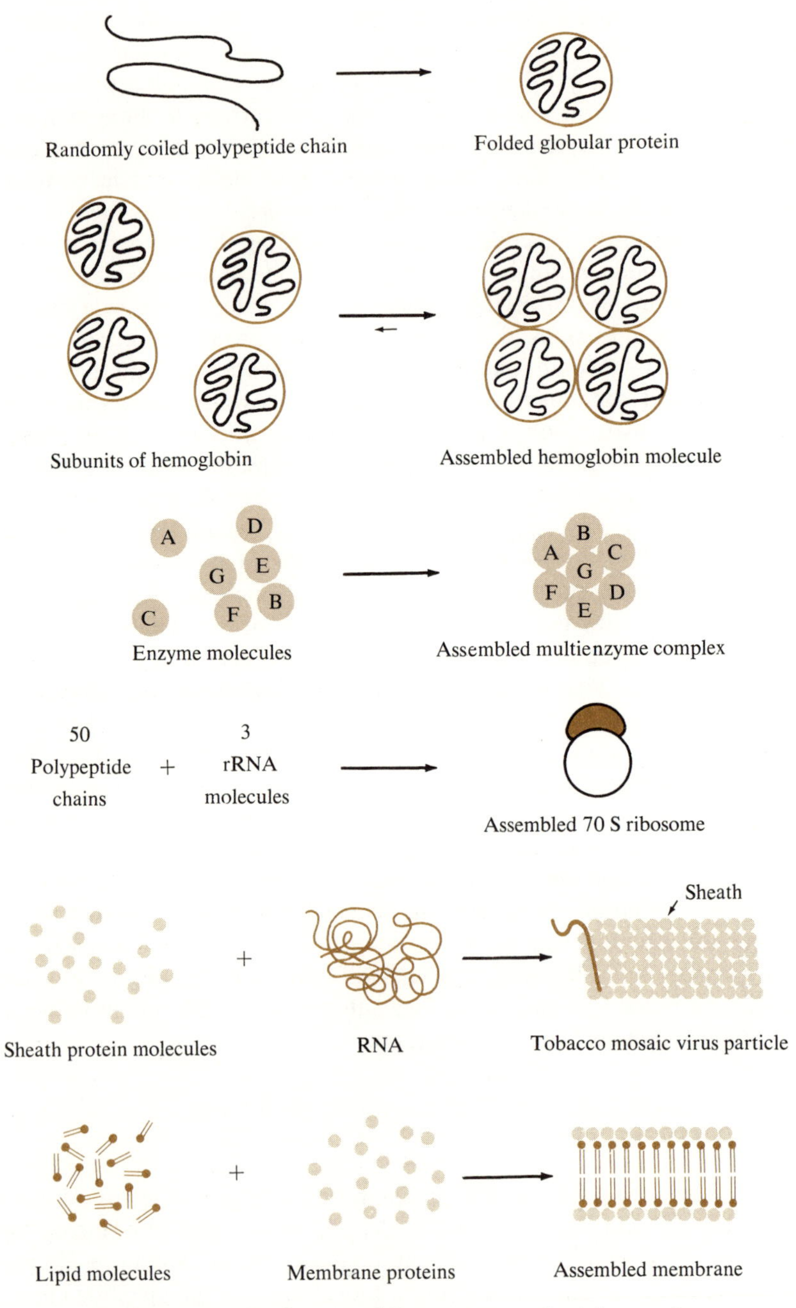

Figure 9–5. Self-assembling supramolecular systems

(with a molecular weight of 64,500) contains four polypeptide chains. Each of the four chains exists in a complex three-dimensional arrangement that is very similar to that of the single chain of myoglobin. In fact, both hemoglobin and myoglobin have the same basic biological function, the transport of bound oxygen molecules. However, the four chains of hemoglobin are never found in the free state in the cell; they are always associated with each other in a cluster of four as the globular hemoglobin molecule. The reason is not hard to find. When the four chains are separated and then mixed, they automatically come together and associate tightly to form complete functional hemoglobin molecules. In Fig. 9–5 we see that an equilibrium exists between the subunits and the complete molecule. But this equilibrium lies very far in the direction of intact hemoglobin molecules. It has been concluded that the amino acid sequence of each of the four polypeptide chains of hemoglobin specifies not only the three-dimensional configuration of each chain taken separately, but also that the four units must combine inevitably to form the complete hemoglobin molecule. Specific surface side chains of each of the four globular chains are programmed through their amino acid sequence to have complementary patches that fit together precisely. Again, the driving force that compels the subunits to stay together is the thermodynamic tendency of the surrounding water molecules to seek the state of maximum entropy, one that does not permit nonpolar surfaces to remain exposed to water. As a consequence "sticky" patches on the subunits tend to face each other and thus become hidden within the complete hemoglobin molecule.

In hemoglobin we have the simplest self-assembling supramolecular system, one consisting of four parts, all polypeptide chains, whose assembly, three-dimensional structure, and biological function are all ultimately conferred on it by a linear sequence of bases in DNA. This simple system is the prototype of yet more complex self-assembling three-dimensional biostructures. For example, the fatty acid synthetase complex of yeast, which is responsible for synthesis of long-chain fatty acids from acetyl CoA, contains seven enzyme molecules, each of which consists of three polypeptide chains, which have been separated. The 21 separate polypeptide chains, when mixed together, spontaneously assemble themselves into a complete functioning three-dimensional multienzyme complex, through the same principles that result in the self assembly of hemoglobin.

Recent research has shown that much larger biostructures may assemble themselves. The major subunit of *E. coli* ribosomes, which contains thirty polypeptide chains and two different RNA molecules, and the minor subunit, which contains twenty polypeptide chains and a single RNA, will assemble themselves spontaneously from their components. These subunits may then combine to yield a biologically active ribosome capable of synthesizing proteins.

Even virus particles may assemble themselves from their components to yield biologically active end products. For example, the tobacco mosaic virus particle (particle weight 40 million) contains a single RNA molecule and 2200 identical polypeptide chain subunits. It is a tubelike structure in which the protein subunits form the wall and the RNA, which is helically coiled, forms the core. The RNA and the polypeptide subunits can be separated in pure form. When these components are simply mixed together, complete virus particles having the capacity to infect host cells will form spontaneously (Fig. 9–5). The very much larger bacterial virus T_4 of *E. coli* (particle weight, 220,000,000) also can spontaneously assemble much of its own structure from its many components. The forces causing the thousands of component subunits of these viruses to come together automatically to yield a stable structure are the same as those holding together the much simpler hemoglobin molecule, namely hydrophobic interactions, which is the name given to the tendency of water molecules to assume a random, high-entropy state that disallows exposure of the nonpolar or greasy parts of molecules to water. The amino acid sequence of the protein subunits of viruses is evidently programmed so that the virus structure is formed as the inevitable, automatic, and unique product because it is thermodynamically the most stable structure.

Recently it has been found that lipid bilayer membranes also tend to form automatically from lipid molecules of the right size and shape (Fig. 9–5), again because of the tendency of surrounding water molecules, which do not "like" the greasy hydrocarbon chains of the lipids, to force the lipid molecules into such a geometrical pattern that the hydrocarbon tails are hidden within the bilayer structure, with only the polar hydrophilic "heads" of the lipid molecules exposed. Such lipid bilayers also can gather to themselves specific protein molecules, whose amino acid sequence determines they will fit the lipid layer in a thermodynamically stable arrangement.

10

ACTIVE TRANSPORT

All cells must expend energy to transport solute molecules against gradients of concentration, either into or out of the cell. This kind of cellular work is not very conspicuous to the naked eye, but we shall see that it is an important and fundamental activity of all cells and requires a large input of metabolic energy. In this chapter we shall first examine the characteristic properties of membranes and the thermodynamic principles underlying membrane transport processes. Then we shall consider the characteristics of active transport systems, particularly those that are present in most cells. Finally, we shall examine some biological specializations of active transport.

10–1 MEMBRANES AND THEIR PERMEABILITY

In order to understand the properties of active transport systems we must first examine the structure, properties, and permeability of cell membranes, which form the boundary between the internal aqueous compartment of the cell and the external environment. Most cell membranes contain about 50 per cent protein and 50 per cent lipid; most of the latter is in the form of phospholipids. Nearly all cell membranes show a high resistance to electrical currents and thus do not allow ions to pass readily; they are good electrical insulators. Under the electron microscope cell membranes show a laminar or layered structure with a total thickness of about 90 Å. In order to account for these properties, Robertson postulated what he called the "unit membrane" hypothesis of membrane structure, shown schematically in Fig. 10–1.

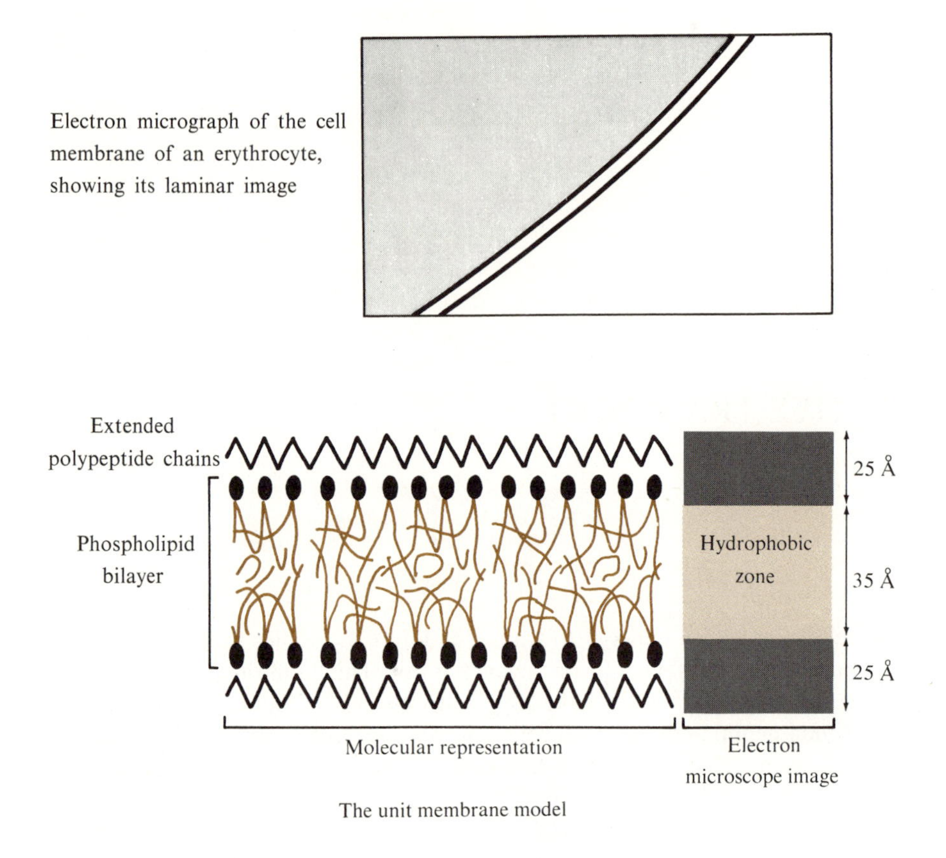

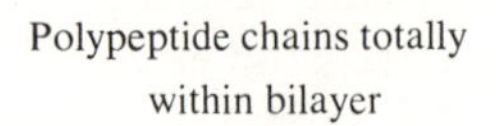

Polypeptide chains totally within bilayer

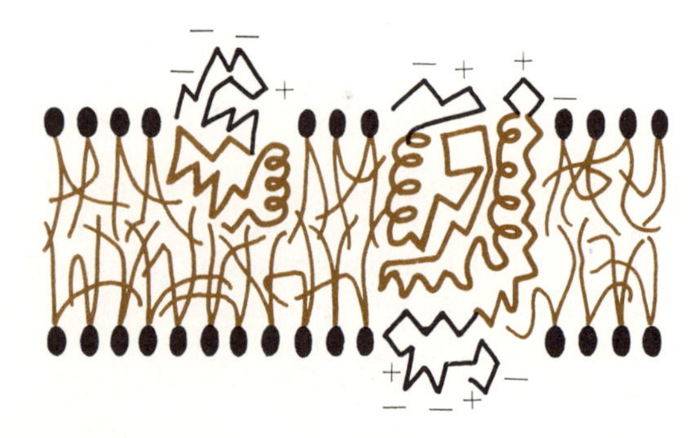

Hydrophobic portions of polypeptide chains within bilayer

Other models of membrane structure

Figure 10–1. Some models of membrane structure.

According to this concept, which was proposed in the 1950's, membranes contain a double layer or *bilayer* of phospholipid molecules (Chapter 7), in which the hydrophobic or water-hating hydrocarbon chains of the lipids face each other to form a continuous hydrocarbon barrier (Chapter 9). Such a hydrocarbon phase is oily in nature and cannot conduct electrical currents. He also proposed that the lipid double layer is coated on each side with a single layer of extended protein molecules.

Although the unit membrane hypothesis explains many aspects of membrane properties, other models of membrane structure have been postulated. One of these models suggests that the polypeptide chains of the membrane protein penetrate into the lipid bilayer, in part or entirely; another proposes that the lipid molecules are arranged in globular, protein-coated units. Whatever their precise structure, cell membranes have great physical strength and high electrical resistance, despite the fact that they are exceedingly thin.

Some generalizations may be made regarding the permeability of cell membranes. In most membranes the rate of penetration of uncharged solutes is proportional to their solubility in nonpolar or oily liquids, such as olive oil. Thus, neutral molecules having hydrocarbon chains, such as aliphatic alcohols (an example is 1-butanol, $CH_3CH_2CH_2CH_2OH$), tend to penetrate membranes very rapidly. Since the inner core of the membrane is itself an oily hydrocarbon layer, we can see that any substance that dissolves in this inner layer might also pass through the membrane readily. (Water is an exception; although it is not soluble in olive oil, it penetrates membranes very readily, presumably along hydrated channels.) However, larger, more polar neutral molecules, such as the sugar sucrose, which has many water-loving hydroxyl groups, pass through most membranes only slowly, in accordance with their low solubility in olive oil, whereas electrically charged molecules, such as the ions Mg^{2+}, Li^{+}, Cl^{-}, and $HPO_4{}^{2-}$, have very little tendency to pass through most membranes since they are highly polar.

From these considerations, it is clear that the structure of membranes is adapted to prevent polar or electrically charged molecules from passing through them freely. Yet there are two well-known facts that at first do not appear to be consistent with this generalization. The first is that cells must continuously exchange large amounts of certain polar molecules with their surroundings. For example, most cells must take in such nutrients as glucose, which has several hydroxyl groups and is thus a polar molecule, and amino acids, which have two or more electrical charges at pH 7.0. Moreover, we know that some electrically charged molecules must leave cells as end products of metabolism, such as lactate and bicarbonate ions. However, these cases are exceptions to the general rule that polar molecules cannot penetrate cell membranes readily. The great majority of biologically occurring cations and anions do not pass through the cell membrane; among them are ADP^{3-}, ATP^{4-}, glucose 6-phosphate^{2-}, as well as the tricarboxylic acid cycle intermediates succinate^{2-}, oxaloacetate^{2-}, and citrate^{3-}. In fact, most of the hundreds of known metabolic intermediates are polar or charged molecules that are locked within the cell because of the impermeability of the cell membrane.

Let us now consider the exceptions. Why is it that only certain polar molecules, such as glucose and amino acids, can cross the cell membrane readily, whereas most polar molecules cannot? The answer is that the cell membrane contains specific macromolecules, believed to be proteins, whose function it is to bind and carry certain polar nutrients or metabolic end products across the membrane; they are variously called transport systems, carriers, porters, or permeases.

There is a second apparently anomalous property associated with the permeability of the cell membrane. Those few polar compounds that can cross the cell membrane readily, such as glucose and amino acids, are not usually found in equal concentrations on the two sides of the membrane, as we might expect if the membrane is freely permeable in both directions. For example, the concentration of amino acids inside the cells of vertebrates is much higher than their concentration in the extracellular fluid or blood. The reason is that the transport systems that facilitate the passage of amino acids or glucose across the membrane are "energized" in some manner to generate a concentration gradient across the membrane. The transport of specific substances such as glucose or amino acids against a concentration gradient is called *active transport*. Before we go any further, let us examine the energetics of transport across membranes more closely.

10–2 PASSIVE AND ACTIVE TRANSPORT

If we have a semipermeable membrane separating two aqueous compartments and add to one of them a solute that can pass through the membrane readily, we will find that the solute will pass into the other compartment and continue to do so until its concentration is precisely the same on both sides of the membrane. The rate of the net movement of the solute will be rapid at first and will gradually slow down until finally equilibrium is reached, at which point there is no further *net* movement of solute in either direction. At this equilibrium point, the rate of transfer of the solute from the first compartment to the second will be exactly counterbalanced by the transfer of solute in the opposite direction. This tendency of solute molecules to distribute themselves evenly across such a semipermeable membrane is of course the result of the operation of the Second Law of thermodynamics, since in this process the entropy of the solute molecules becomes maximized as they randomize themselves by diffusion throughout the two compartments; simultaneously, the free energy of the system decreases to a minimum. Such a movement of solute molecules across a semipermeable membrane, *down* a

Passive transport

(*down* a concentration gradient)

Semipermeable

Initial state

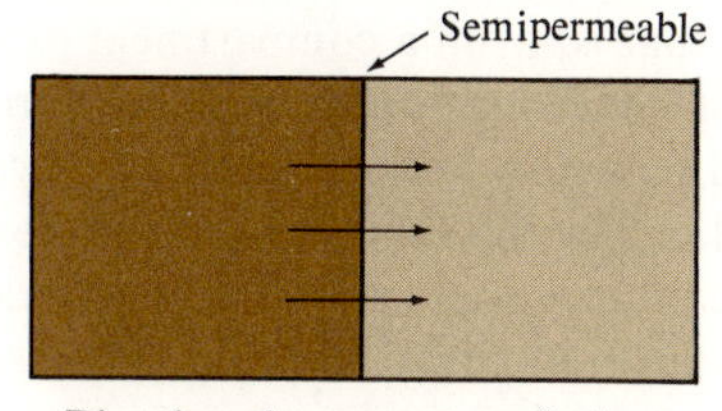

Direction of net transport of solute

The free energy of the system decreases.

Final state (equilibrium)

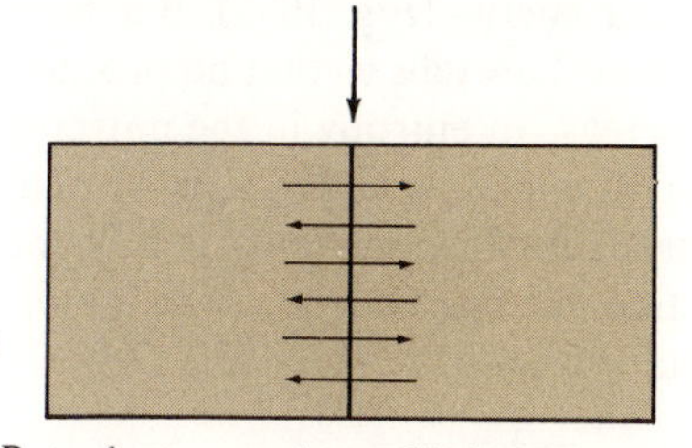

Rate of movement is equal in both directions

Active transport

(*against* a concentration gradient)

Initial state

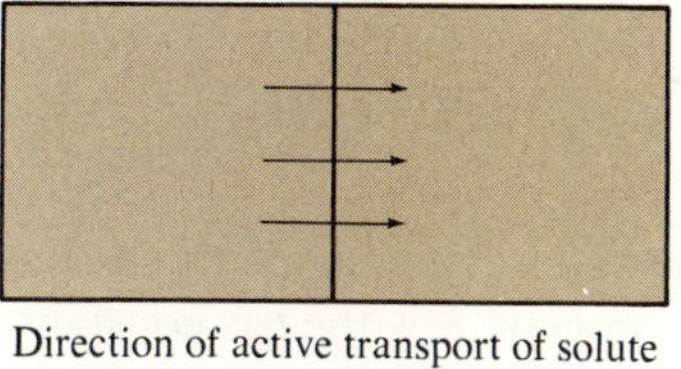

Direction of active transport of solute

The free energy of the system increases. Active transport cannot occur without the input of free energy.

Final state

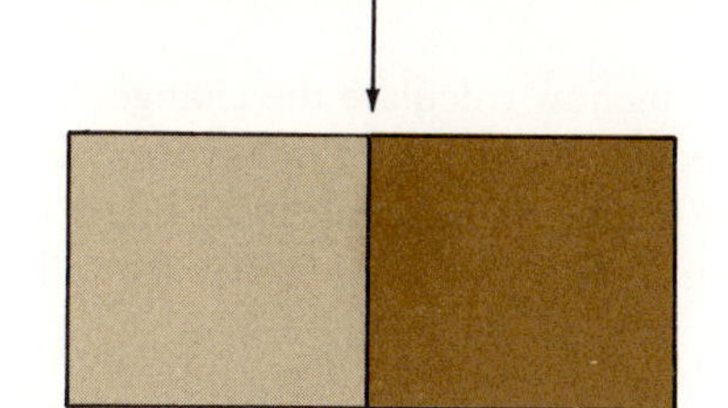

Steady-state gradient

Figure 10–2. Schematic representation of passive and active transport. The density of color in the chambers is proportional to solute concentration.

concentration gradient, toward the position of equilibrium, is called *passive transport*.

Now let us see the consequences if a solute should move against or up a concentration gradient, that is, from a compartment of low concentration to a compartment of high concentration, as occurs during active transport. Such a movement would obviously decrease entropy, since the solute molecules would become less random; by the same token, the free energy of the system would increase. We therefore have a very simple criterion for distinguishing active and passive transport: an active transport process is one in which the system gains in free energy and a passive transport process is one in which the system decreases in free energy (Fig. 10–2). But here we come to a very important point. The Second Law tells us that no process can occur spontaneously that results in a decrease of entropy in the universe or in a net increase in free energy of the system under consideration. An active transport process thus cannot occur by itself; it can occur only if it is coupled to some other process that can yield free energy. On the other hand, a passive transport process can occur spontaneously.

10-3 THE ENERGY REQUIREMENT OF ACTIVE TRANSPORT

We can calculate the change of free energy when 1.0 mole of an uncharged solute is moved from one compartment to another, if we know the concentration of the free solute in each compartment. The basic relationship is given by the following equation

$$AG^0 = 2.3\ RT \log_{10} \frac{C_2}{C_1}$$

where C_1 and C_2 are the concentrations of the free solute at the beginning and end of the transport process, R is the gas constant, and T is the absolute temperature. This equation assumes that the solute molecule has no net charge.

With this equation let us now calculate the change in free energy in transporting one gram molecular weight of glucose up a hundredfold gradient from a compartment in which its concentration is, let us say, 0.001 M to a compartment in which its concentration is 0.1 M. We will have

$$\begin{aligned} \Delta G^0 &= 1.98 \times 298 \times 2.3 \times \log_{10} \frac{0.100}{0.001} \\ &= 1.98 \times 293 \times 2.3 \times 2.0 \\ &= 2680 \text{ cal} \\ &= 2.68 \text{ kcal} \end{aligned}$$

Since the free energy change is positive in sign this process is one of active transport; it is endergonic and cannot proceed spontaneously. In order for it to go to completion at least 2.68 kcal of free energy must be applied to the system for each mole of glucose transported.

Such thermodynamic calculations give us only a minimum statement of how much free energy a cell must invest to transport a given solute against a given gradient. Usually much more is expended. Moreover, as we shall see, there is always back leakage or efflux of the solute, which opposes the active pumping process. If the rate of back leakage is high, the pump has to work harder to maintain a given gradient. The gradients of solutes that are observed across the membranes of living cells are thus the result of dynamic steady states.

If we should now calculate the free energy change for the passage of glucose from a compartment in which its concentration is 0.1 M to another in which its concentration is 0.00 M, that is, *down* a concentration gradient of $100 \rightarrow 1$, the free energy change will be the same in magnitude but negative in sign, (i.e., -2.68 kcal), indicating that the process is passive and may occur spontaneously.

The calculations we have just made apply only to uncharged solute molecules. When an electrically charged molecule or ion, such as Na^+, undergoes uphill transport, there are two gradients against which the ion is moved. One is a gradient of concentration and the other is a gradient of electrical charge. The sum of these two gradients is called the *electrochemical gradient.* More work is required to move a charged ion up an electrochemical gradient than an uncharged molecule.

The concentration gradients across the membranes of cells vary widely and may be as high as 10,000,000 to 1.0. One of the largest gradients known occurs in the parietal cells present in the epithelial cell layer lining the stomach of man and other mammals, which secrete hydrochloric acid into the gastric juice to concentrations exceeding 1.0 M. The HCl is secreted by cells whose intracellular hydrogen ion concentration is believed to be only about 0.0000001 or 10^{-7} M. Thus parietal cells can move H^+ ions up a gradient of about $1 \rightarrow 10{,}000{,}000$, indicating not only that these cells must have very efficient "pumps" for secreting H^+ ions, but also that considerable energy is required to achieve this degree of concentrative work.

10-4 CHARACTERISTICS OF ACTIVE TRANSPORT SYSTEMS

To illustrate the important features of active transport systems of membranes let us consider the red blood cells of certain mammals. Chemical analysis has demonstrated that the ionic composition of the internal aqueous phase of the human red blood cell is quite different from that of the blood plasma

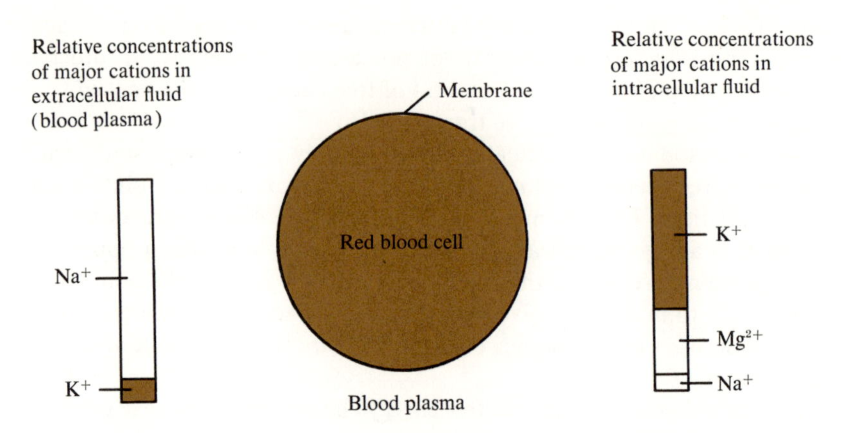

Figure 10–3. Gradients of K^+ and Na^+ across the cell membrane of the erythrocyte. They are maintained by energy-dependent transport of Na^+ out of the cell and K^+ into the cell.

which bathes it. As Fig. 10–3 shows, the intracellular aqueous compartment is high in K^+ ions but contains relatively little Na^+, whereas the extracellular blood plasma is rich in Na^+ and contains very little K^+. It might be thought that the existence of these gradients in the concentrations of K^+ and Na^+ across the red cell membrane are merely reflections of the inability of these ions to pass through the membrane, thus preventing their attaining equilibrium. This is not the case; the red cell membrane is in fact significantly permeable to both K^+ and Na^+. Tracer experiments with radioactive K^+ and Na^+ have shown that K^+ is actively pumped into the red blood cell, whereas Na^+ is actively pumped out. However, both ions tend to leak back again. The result is a steady state in which the internal K^+ concentration normally remains high and constant with time.

The first identifying characteristic of an active transport process is that it depends on a source of metabolic energy to pump a solute against a gradient of concentration. That the characteristic distribution of Na^+ and K^+ across the membrane of the red blood cell requires energy can be shown by a simple experiment. Mature red blood cells of mammals obtain all their energy from the glycolytic conversion of glucose into two molecules of lactate, a process that we have seen is accompanied by the coupled formation of two molecules of ATP. As long as glycolysis and ATP production take place, the characteristically high internal K^+ concentration will be maintained at a constant level. However, if we should stop glycolysis (and thus prevent ATP production) by adding fluoride, which inhibits the glycolytic enzyme *enolase*, we will find that the internal concentration of K^+ will gradually fall and that of Na^+ will rise until a point is reached at which the concentrations of K^+ and Na^+

become equal on both sides of the membrane. Clearly the maintenance of the characteristic Na^+ and K^+ concentrations inside the red blood cell is the result of an energy-dependent process, presumably one requiring ATP. In other kinds of cells such as liver or kidney cells, oxidative phosphorylation is the main source of ATP energy required to support active transport mechanisms. In such cells respiratory inhibitors, such as cyanide, or uncouplers of oxidative phosphorylation, such as 2,4-dinitrophenol, will inhibit active transport processes.

A second important property of the active transport systems of cells is that they are specific for given solutes. Some cells have a "pump" specific for certain amino acids, but cannot transport glucose. Other cells may have a pump for glucose but none for amino acids. In fact, active transport systems have considerable specificity, resembling that of enzymes for their substrates. For example, the red blood cells of some mammals transport D-glucose inward at a very high rate, but can transport D-fructose only very slowly. Moreover, other types of cells have a pump specific for amino acids with neutral or uncharged R groups such as glycine and alanine, but cannot transport amino acids whose R groups have an electrical charge, such as glutamic acid or lysine. (We shall often use the expressive term "pump" in referring to biological active transport mechanisms. However, no similarity of design or mechanism to a man-made pump is intended; the molecular mechanisms by which active transport is brought about do not physically pump or expel the transported substance.)

Active transport systems have a third characteristic property, one that also is highly reminiscent of enzyme action. Their activity depends on the concentration of the substance being transported. For example, when glucose is actively transported into a cell, the rate of glucose influx increases with the external concentration of glucose. However, a characteristic plateau is soon reached in the rate of glucose influx so that any further increase in the external concentration of glucose will produce no corresponding increase in glucose influx (Fig. 10–4). Thus an active transport system can be "saturated" with the substance transported, just as an enzyme can be saturated with its substrate.

A fourth characteristic of active transport systems is that they have a specific directionality. In most cells K^+ is pumped only in the inward direction. Moreover, the specific pumps for glucose and for amino acids are also directed inward. On the other hand, Na^+ is usually pumped outward. Active transport is thus a *vectorial* or directional process.

A fifth important property of active transport systems is that they may be selectively poisoned. For example, active transport of glucose in the kidney is poisoned by the substance *phlorizin*, a glycoside from the bark of the pear tree. The active transport of Na^+ out of the erythrocyte is inhibited by the toxic glucoside *ouabain*. This property may remind us of the poisoning of enzymes by specific inhibitors.

Finally, we may note that the integrated action of active transport mechanisms can maintain the internal solute and ion composition of cells at a remarkably constant level even when the external medium fluctuates widely

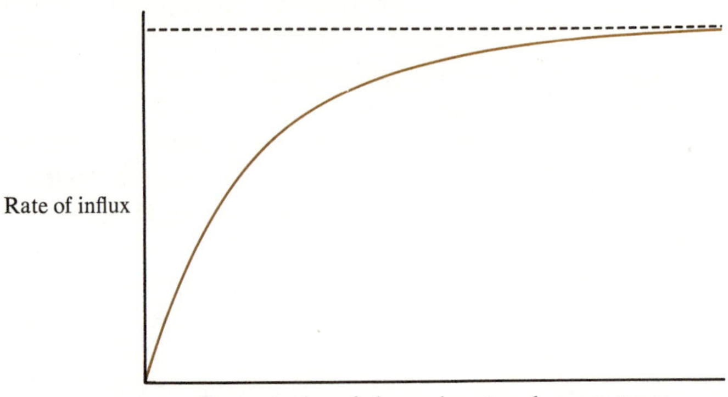

Figure 10–4. Saturation of the glucose transport system at high glucose concentrations.

in composition. This attribute is particularly noteworthy in unicellular organisms, such as yeasts and bacteria, which have remarkable ability to cope with wide fluctuations in the composition of the external milieu. Yeast cells can live and grow and thus keep their internal solute composition compatible with cellular function when the pH of the culture medium is varied from pH 3 to pH 10—that is, a 10 millionfold change in H^+ ion concentration. Yeast cells also can survive large variations in the Na^+ and K^+ content of the medium. Thus the active transport systems in the membranes of cells must be able to sense the relative concentrations of their specific substrates or *ligands* in the internal and external compartments and adjust their pumping rates accordingly. In many respects active transport systems in cell membranes resemble enzymes; they show substrate specificity, they can be inhibited, and they can be saturated by their substrates.

Active transport systems are believed to contain two major components (Fig. 10–5). The first is a protein molecule with a specific binding site for the substance transported, resembling the substrate binding site of an enzyme. It is the function of this component, called the carrier or porter, to translocate the ligand across the membrane and "unload" it on the other side. The second component, presumably a protein or group of proteins, has the function of transferring energy to the carrier, possibly in the form of a high-energy covalent bond, to enable it to carry the substrate against a gradient of concentration.

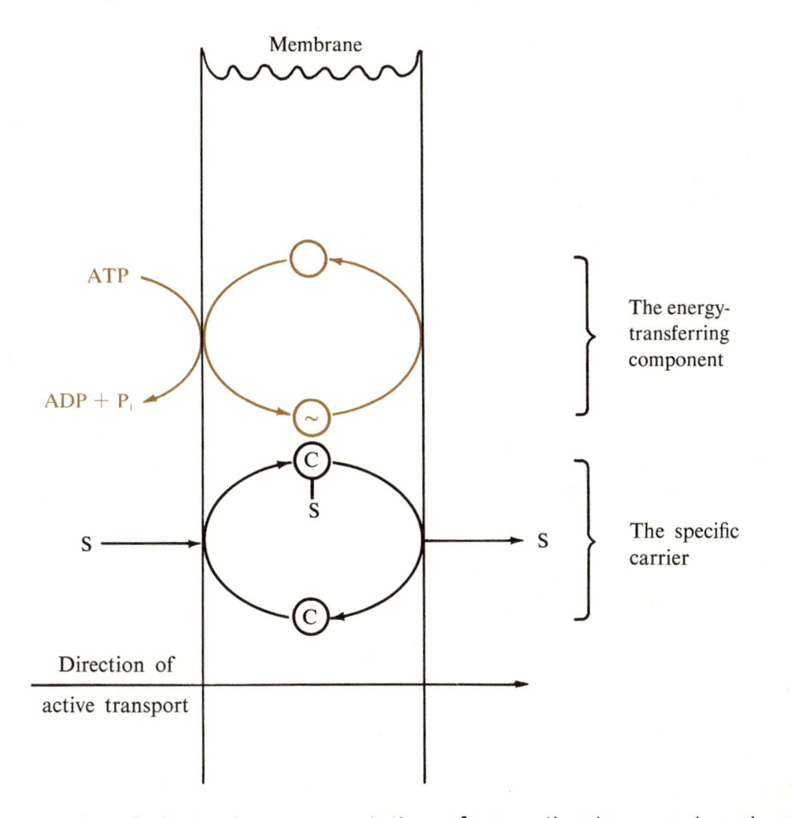

Figure 10–5. Schematic representation of an active transport system. The energy-transferring process (color) makes possible transport of S against a gradient.

10–5 THE MAJOR ACTIVE TRANSPORT SYSTEMS OF CELLS. THE TRANSPORT OF K^+ AND Na^+

There are two general types of active transport systems or pumps, which nearly all cells possess. One type is concerned with maintaining a proper balance of K^+, Na^+, and water in the cell. The other set is concerned with the inward transport of essential organic nutrients, in particular, glucose and amino acids. In addition to these major pumps there are others that are specialized for secondary or auxiliary transport functions.

Nearly all cells of vertebrates, and not only red blood cells, contain a relatively high and constant internal concentration of K^+, about 100 to 150 mM, and a very low concentration of Na^+; this condition is true also of many bacterial and plant cells. A high internal concentration of K^+ is necessary for maximum activity of certain vital enzymes, in particular for pyruvate kinase, one of the enzymes of glycolysis; K^+ is required also for maximal rates of protein synthesis by ribosomes. Na^+ cannot replace K^+ in these processes. Most cells of higher animals have an active transport system that pumps Na^+ out of the cell, against a concentration gradient, and pumps K^+ into the cell, also against a gradient. The transport of both Na^+ and K^+ is inhibited by ouabain.

Our present knowledge of the molecular components of the pump for Na^+ and K^+ stems from a discovery made some years ago. Fragments of the membranes of crab nerve cells contain an enzyme that hydrolyzes ATP to form ADP and phosphate. This enzyme was found to be stimulated when *both* Na^+ and K^+ were added. Moreover, it was discovered that ouabain inhibits that portion of the hydrolysis of ATP that is stimulated by $Na^+ + K^+$. From these observations it was concluded that this ATPase must be responsible for the active transport of Na^+ and K^+ across the cell membrane. This enzyme, called the Na^+-, K^+-dependent ATPase, has since been found in the plasma membrane of virtually all cells in vertebrate tissues and it is now accepted as the primary transport system by which Na^+ is pumped out of cells and K^+ pumped in.

As a result of many recent investigations, it has been concluded that the Na^+-, K^+-dependent ATPase system, which has a large particle weight and probably consists of two or more component protein molecules, is fixed in the cell membrane in such a way that it carries out unidirectional transport of Na^+ out of the cell and K^+ into the cell. For each molecule of ATP hydrolyzed, three Na^+ ions are expelled from the cell and nearly an equal number of K^+ ions move inward; a direct and quantitative coupling between ATP hydrolysis and the transport process therefore exists. Presumably the free energy decrease that occurs when ATP undergoes hydrolysis to ADP and phosphate is used to cause a configurational change, or possibly a rotation of the ATPase molecule in the membrane, so that Na^+ is caused to move out and K^+ in. Recent research indicates that there is a chemical intermediate in this process, since a functional group on the Na^+-, K^+-stimulated ATPase molecule has been found to undergo phosphorylation by transfer of the terminal phosphate group from ATP, before the enzyme completes the transport of Na^+ and K^+. The ATP required to energize the transport process, as well as its products ADP and phosphate, remain within the cell at all times. These relationships are summarized in Fig. 10–6.

In those cells that are very active in secretory or transport processes the Na^+-, K^+-stimulated ATPase of the cell membrane may use a major fraction of the cell's ATP output. For example, the epithelial cells of the kidney, which secrete Na^+ ions from the blood into the urine and which reabsorb K^+ from the blood, utilize over two-thirds of the total ATP output of their mitochondria to pump Na^+. In the brain an even larger fraction of the cell's output of ATP is used by the Na^+-, K^+-stimulated ATPase. The active transport of Na^+ and K^+ across the membrane of nerve cells is an important

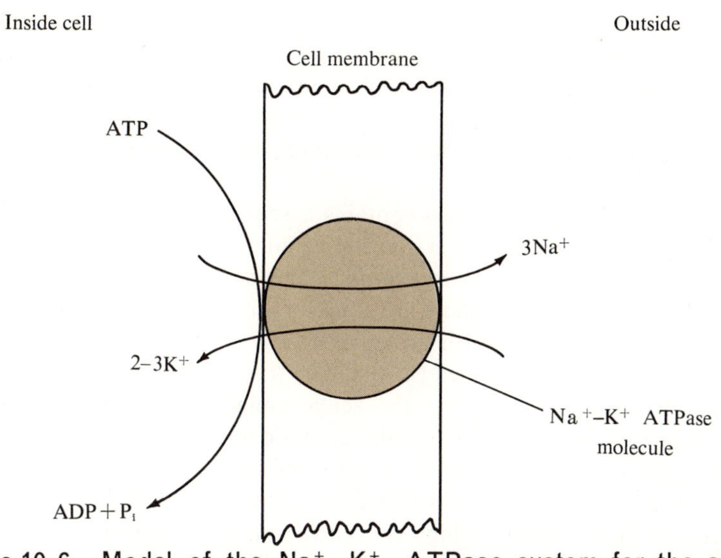

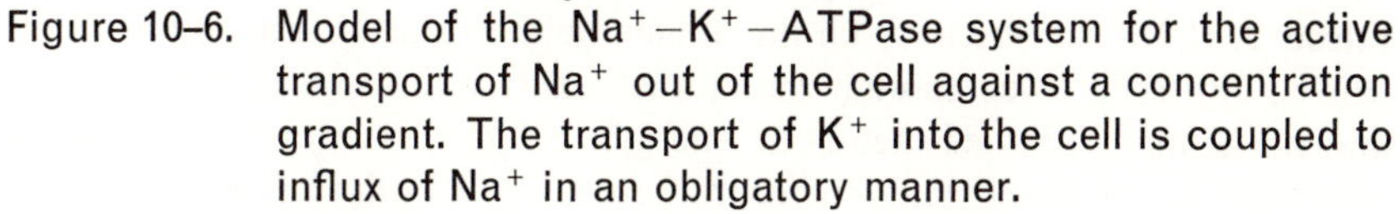
Figure 10–6. Model of the Na^+–K^+–ATPase system for the active transport of Na^+ out of the cell against a concentration gradient. The transport of K^+ into the cell is coupled to influx of Na^+ in an obligatory manner.

process enabling them to recover following propagation of a nerve impulse in the form of an action potential, as we shall see later.

The Na^+-, K^+-stimulated ATPase in the cell membrane aids also in maintaining a proper water balance in the cell. If K^+ were continuously pumped into the cell, without compensating loss of a cation from the cell to the outside, water would necessarily flow into the cell with the K^+ and would cause it to swell. By pumping an equal amount of Na^+ out of the cell, and with it an equivalent amount of water, the cell can maintain not only the proper internal concentration of K^+ but also the proper water balance.

10–6 GLUCOSE AND AMINO ACID TRANSPORT

The second major type of active transport system is required to transport vital organic nutrients into the cell from the surrounding medium, where they may be very dilute. Such nutrients may be required as a source of energy, as in the case of glucose, or a source of nitrogen, as in the case of amino acids. Such nutrient transport systems are particularly well developed in bacteria and in the epithelial cells of the vertebrate small intestine, which are very active in absorbing nutrients into the bloodstream.

For example, cells in the epithelium of the small intestine can absorb certain hexose sugars, such as glucose, fructose, galactose, and mannose, with a high degree of specificity. This process has been found to be the result of inward-directed active transport, since glucose can be absorbed from very dilute solutions placed in the lumen of the intestine, which are far lower in glucose concentration than the cytoplasm of the epithelial cells or the blood draining the intestine. Moreover, it has been established that glucose absorption is dependent upon respiration and thus requires energy, presumably in the form of ATP.

The intestinal epithelium also contains a series of pumps for the absorption of free amino acids from the lumen of the intestine. These systems likewise require energy and transport amino acids into the cell against a gradient of concentration. There are at least four different amino acid pumps, each of which is specific for a group of closely related amino acids. One is specific for neutral amino acids, and can transport glycine, alanine, and leucine. Another is specific for the acidic amino acids glutamic acid and aspartic acid. Still others are specific for basic amino acids and for proline.

The active transport systems for sugars and amino acids in the epithelium of the small intestine share an interesting and highly significant property: they require the presence of high concentrations of Na^+ in the lumen of the intestine. If Na^+ is absent from the lumen, neither glucose nor amino acids can be absorbed, even though the epithelium is well supplied with oxygen and nutrients to provide metabolic energy. In fact, for inward transport of either glucose or amino acids to occur, there must be both an inward gradient of Na^+ and inward transport of Na^+. Since we have already seen that animal cells normally pump Na^+ outward against a gradient, we may well ask why it is that Na^+ moves into the cell along with glucose or amino acids.

A rather interesting hypothesis to account for these observations (Fig. 10–7) proposes that the active energy-linked transport of glucose into the cell requires the cooperation of two transport systems. The first is a carrier or translocase capable of binding glucose reversibly and transporting it in either direction across the membrane. However, this carrier can translocate glucose only if Na^+ is simultaneously transported in the same direction. It is postulated that both glucose and Na^+ are bound to the same carrier, but at two different sites; after they are carried into the cell they dissociate from the carrier into the cytoplasm. The second major component required for active transport of glucose into the cell is an outwardly directed Na^+ pump that is driven by hydrolysis of ATP. This pump does not combine with or transport glucose; its sole purpose is to transport Na^+ out of the cell at the expense of the energy yielded on hydrolysis of ATP, thus yielding an inward gradient of Na^+ ions.

These two transport systems cooperate in the following manner to bring about active transport of glucose into the cell against a concentration gradient. The outwardly directed ATP-dependent Na^+ pump continuously maintains a Na^+ gradient across the cell membrane, so that at all times the external Na^+ concentration is much higher than the internal. The inward gradient

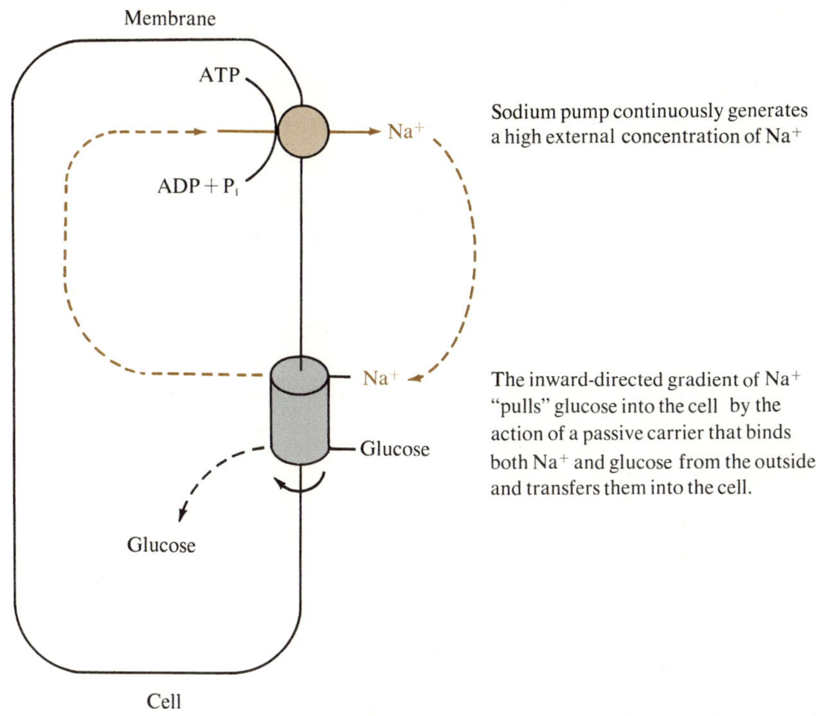

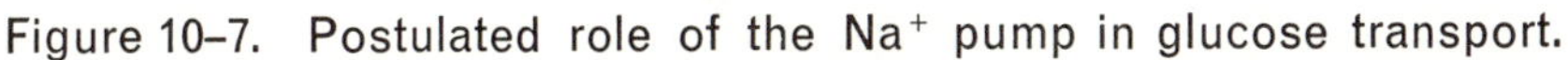
Figure 10–7. Postulated role of the Na^+ pump in glucose transport.

of Na^+ so generated then serves as the driving force to "pull" glucose into the cell. This is accomplished by the binding of both external Na^+ and external glucose to the glucose carrier molecule, which we have seen cannot transport glucose without Na^+. The binding of Na^+ ions to the glucose carrier and the tendency of Na^+ to move inward down its gradient is the force that ultimately pulls glucose into the cell. Inside the cell both Na^+ and glucose leave the carrier and the Na^+ is pumped out again. Naturally, the outward gradient of glucose can never be greater than the inward gradient of Na^+ generated by the Na^+ pump. Thus the pumping of Na^+ out of the cell is the ultimate driving force for pumping glucose into the cell. Moreover, amino acids also appear to be actively transported into epithelial cells of the intestine by "riding" down an inward gradient of Na^+ ions generated by the outwardly directed Na^+ pump.

From these considerations the more general hypothesis has been formulated that perhaps all inwardly directed active transport processes are coupled to the outwardly directed Na^+ pump by the action of carriers or translocases capable of passively and simultaneously carrying both Na^+ and the specific substance to be transported, whether it be glucose, amino acids, K^+, or some other solute. This basic Na^+ pump has been referred to as the *electrogenic* Na^+ pump, because it can generate an electrochemical gradient. In bacteria, which do not require Na^+ and which do not appear to possess a Na^+ pump, a comparable function may be provided by an outwardly directed pump for H^+ ions.

10–7 THE ORGANIZATION OF TRANSPORT ACTIVITIES IN CELLS AND TISSUES

We may now summarize the ways in which transport processes are geometrically organized in cells and tissues. Three types of organization are recognized: *homocellular*, *intracellular*, and *transcellular* transport.

All cells carry out *homocellular* transport (Fig. 10–8). This type of transport

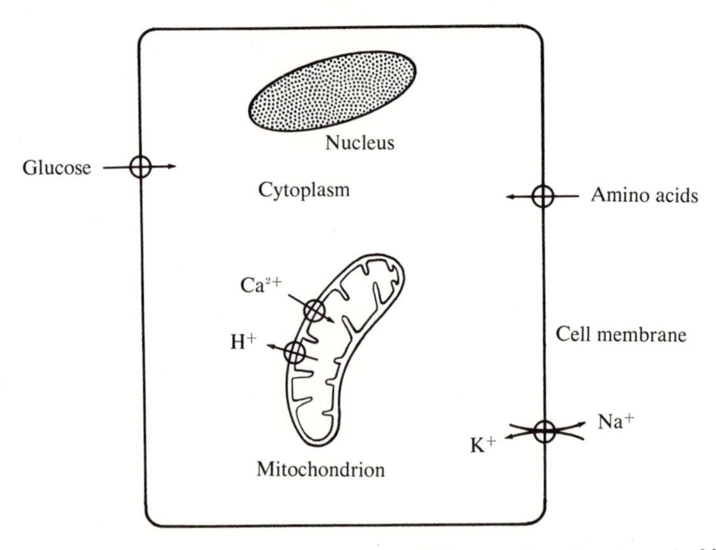

Figure 10–8. Homocellular and intracellular active transport. Homocellular transport maintains the solute composition of the cytoplasm; intracellular transport maintains the solute composition of the compartments within organelles.

process, which takes place in the cell membrane, maintains the concentrations of internal cytoplasmic solutes at optimal steady-state concentrations and thus protects the cell against deleterious effects of fluctuations in the solute content of the external environment. Homocellular transport is especially important in free-living unicellular organisms that may be exposed to wide variations in the composition of the surrounding medium. For example, it allows bacteria to extract essential nutrients from the environment even when their concentration is vanishingly small. In homocellular active transport the molecular pumps are presumed to be distributed more or less uniformly over the entire membrane surface of the cell.

The second general type of transport process is *intracellular transport*, which takes place across the membranes of internal organelles, such as mitochondria, chloroplasts, and endoplasmic reticulum (Fig. 10–8). We have seen that mitochondria possess specific transport systems in their inner membrane, which cause respiration-dependent extrusion of H^+ ions into the surrounding cytoplasm, as they accumulate such ions as Ca^{2+}, from the cytoplasm. Similarly, chloroplasts of green plant cells are capable of accumulating various ions from and secreting others into the cytoplasm at the expense of energy provided by photosynthetic electron flow. Moreover, the membrane of the endoplasmic reticulum of some muscles can rapidly segregate Ca^{2+} ions from the cytoplasm of muscle, in an ATP-dependent reaction that is essential for the relaxation of contracted muscle fibers (Chapter 11). Presumably intracellular transport processes taking place across the membranes of organelles function to maintain the intraorganellar solute composition at optimal levels and to bring about the reversible segregation of certain ions participating in cyclic cellular phenomena, such as occur in the function of muscle, nerve, and secretory cells.

Finally, we have *transcellular* transport processes, which are organized to operate across an entire cell or layer of cells, particularly the epithelial cell layers that line the gastrointestinal tract, the urinary system in vertebrates, and certain glands. In transcellular active transport the whole cell layer constitutes a semipermeable barrier, across which solute gradients are maintained. For example, a very large gradient of H^+ ions is maintained across the parietal cell layer in the stomach epithelium, so that the gastric juice on one side of the cell has a very high concentration of H^+ions (pH $\cong$ 1.0) and the interstitial fluid and blood on the other side of the cell is neutral (pH $\cong$ 7.4 in mammals). In transcellular transport the molecular transport systems are asymmetrically distributed in the cell membrane, so that the portion of the cell membrane on the luminal (mucosal) side of the cell layer is not identical in its transport function with the membrane of the capillary or serosal side of the cell layer (Fig. 10–9). The H^+ ions that are secreted into gastric juice ultimately come from the blood (pH 7.4) and are caused to traverse the parietal cell whose internal pH is probably about 6.0. In addition to effecting this type of directional or asymmetric transcellular transport the parietal cell must keep its own internal K^+ high and constant in relation to both the blood and the gastric juice.

10–8 THE BIOELECTRICAL EFFECTS OF ACTIVE TRANSPORT: ACTION POTENTIALS

The conversion of the chemical energy of ATP into the work of transporting Na^+ and K^+ across cell membranes has another consequence that is exploited

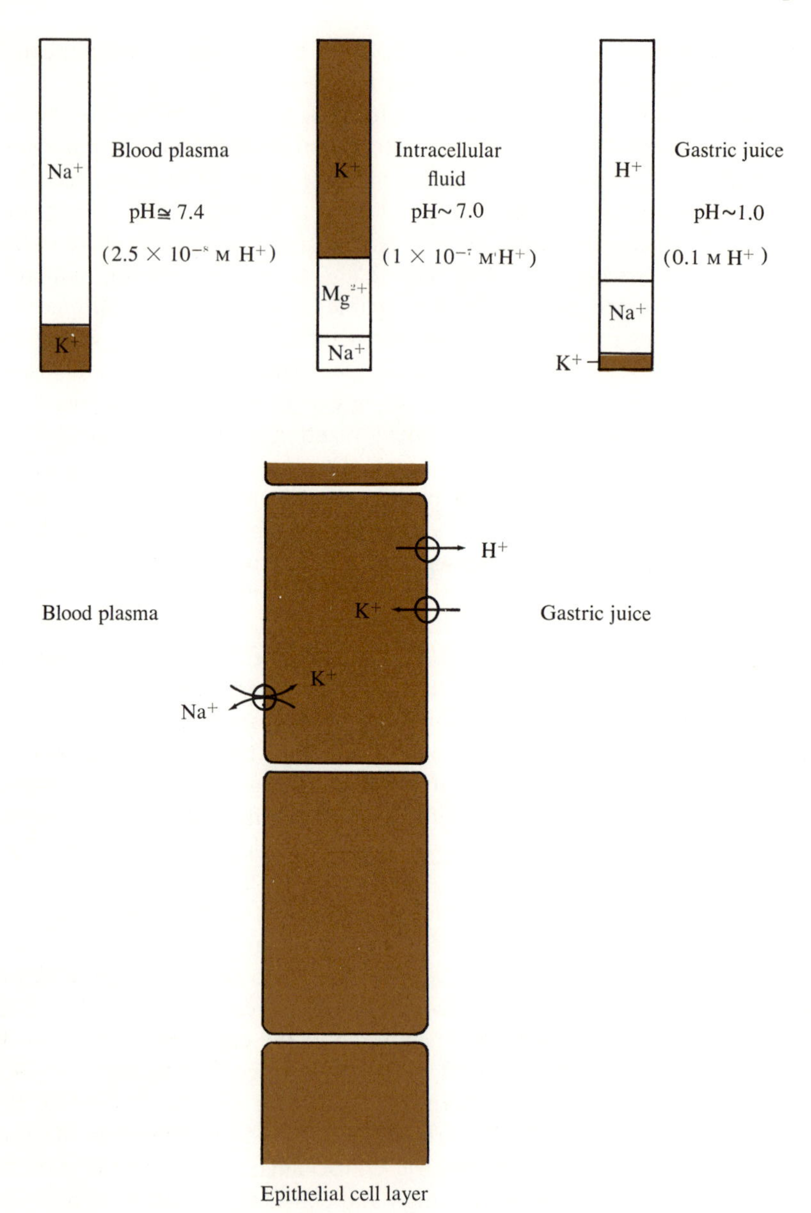

Figure 10–9. Transcellular transport across epithelial cell layer in formation of gastric juice.

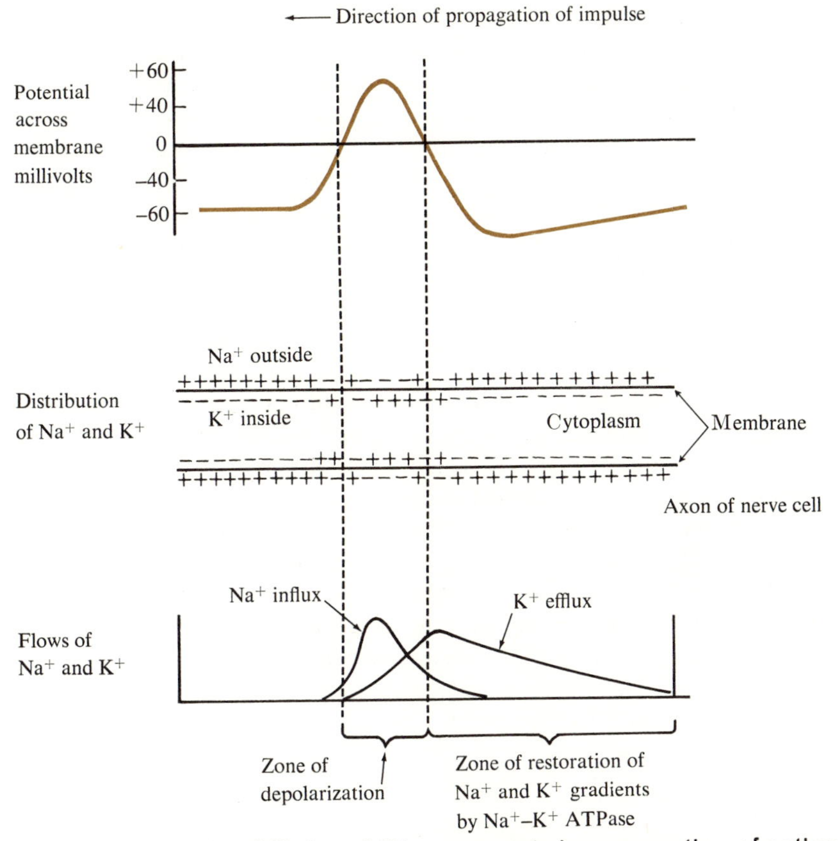

Figure 10–10. Role of Na^+ and K^+ movements in propagation of action potential by axon.

to provide a means of transmitting impulses along nerve cells. In resting nerve cells the gradient of K^+ and Na^+ established across the membrane by action of the ATP-driven pump is such as to yield a difference in electrical potential across the membrane, the inside of the membrane being negative with respect to the outside compartment. The magnitude of this potential difference is a function of the ratio of the concentrations of the charged ions on opposite sides. Thus K^+ is ten to twenty times more concentrated near the inner surface of the membrane, and similarly Na^+ is much more concentrated in the outer compartment. Na^+ and K^+ are sufficiently different from each other with respect to their ionic radius, and hence their charge density, so that in the dynamic steady state (the result of active pumping and passive back diffusion across the membrane) there is a very slight excess of negative charge inside the cell membrane. This potential difference is about 50 to 70 mv, depending on the nerve.

When the nerve cell is excited by reception of an impulse from the preceding cell, or by direct electrical stimulation, an excitation wave travels along the axon at a high rate. This wave is detected by an abrupt change in the potential difference across the membrane and its rapid return to the initial state, which can be measured and recorded from electrodes placed inside and outside the cell. This abrupt change and return of the transmembrane potential is called the *action potential*, and its velocity along the axon may exceed 100 meters/sec, far higher than the rate of diffusion of small molecules.

The action potential is the result of a rapid change of the distribution of Na^+ and K^+ across the membrane (Fig. 10–10). One theory is that local excitation of the membrane at one point causes the lipid and protein molecules of the membrane to reorient transiently in such a manner that the transport systems allow Na^+ and K^+ to flow very rapidly across the membrane to discharge their gradients and thus to cancel out the potential difference across the membrane. In fact, the sign of the transmembrane potential usually reverses, a phenomenon attributed to the unequal rates at which Na^+ and K^+ flow across the membrane. The membrane at this point is spoken of as being *depolarized.* This zone of depolarization travels rapidly along the membrane, like a burning fuse, but in its wake the normal electrical potential difference is quickly restored, presumably aided by the action of the Na^+ and K^+ pumps, to prepare the axon of the nerve cell for the next impulse. Such a mechanism clearly requires a high degree of geometrical and kinetic specialization of the Na^+-, K^+-stimulated ATPase, which is exceedingly rich in cells of the nervous system.

11

CONTRACTION AND MOTION

We shall now discuss the utilization of the chemical energy of ATP for the performance of mechanical work by living cells. Mechanical work is the most visible and most easily measured form of work. For example, the work done by a muscle can easily be measured by determining the distance a given weight is lifted or by the tension the muscle produces. However, the direct conversion of chemical to mechanical energy is unique to living organisms. Skeletal muscle is really a mechanochemical engine that has no familiar counterparts; man-made engines that perform mechanical work are usually driven by heat or electricity.

11–1 TYPES OF CONTRACTILE AND MOTILE SYSTEMS

The best-known biological systems for the conversion of chemical into mechanical energy are those of skeletal muscle cells, which show many variations in structure and function. At one extreme of biological specialization are the flight muscles of flying insects, which perform mechanical work at enormously high rates; they may have contraction-relaxation cycles of a frequency as high as 1000 per second. On the other hand, skeletal muscles of the tortoise or the sloth contract and relax very slowly. In addition to skeletal muscle there is heart muscle, which undergoes periodic wave-like contraction and relaxation in response to stimuli sent out by pacemaker cells, and smooth or involuntary muscle, such as that in the intestinal wall, which has a slow, generalized contractile action. Still other muscles, called catch muscles, can lock in the concentrated state; an example is the adductor muscle of the clam, which can hold its shell closed for long periods without expenditure of energy.

There are other types of biological systems that do mechanical work. The locomotion of amoebae is caused by contractile processes in the cytoplasm. Many paramecia have contractile structures in their cell membrane called *trichocysts*, which are devices for forcible ejection of waste products. Certain bacteria, as well as the spermatozoa of higher animals, possess motile appendages called *flagella*, whose function it is to propel the cell. Other types of cells, more fixed in location, have similar appendages called *cilia*, which move materials past them.

Actually, nearly all cells of higher organisms show the capacity to carry out contractile, torsional, or translocational work of one kind or another, specialized for different purposes. During mitosis and cell division, contractile fibers in the cytoplasm aid in pulling apart the chromosomal material of the nucleus into two zones and in separating the parent cell into its two progeny. Contractile fibers of the cytoplasm serve also an important organizing function during cell development and differentiation. Moreover, the cytoplasm in the axons of some nerve cells is propelled toward the nerve ending by a mechanochemical process.

But the best understood contractile system is that of skeletal muscle. Regardless of the many biological variations in form and function of different types of skeletal muscle cells, all of them possess one basic kind of molecular engine, the *actomyosin system*, which can convert the chemical energy of ATP into mechanical energy. We shall now examine this system in some detail.

11–2 THE DYNAMIC PROPERTIES OF SKELETAL MUSCLE

First let us examine some general properties in the action of muscle that have been deduced from biophysical studies of intact muscle or isolated muscle fibers stimulated artificially by electrical shocks. From experiments in which the rate of contraction, the amount of shortening, and the tension produced have been measured, it has been found that most muscles show a fundamental similarity in the relationship between maximum tension produced and the velocity of the contraction. Furthermore, it has also been found that heat is produced during the contractile cycle of muscles in a very characteristic manner. All muscles at rest give off heat at a low and constant rate, called the *resting heat*. This is the heat given off by the normal metabolic processes in the muscle cells at the "idling" level. When the muscle is stimulated by an electrical shock or by a nerve impulse, a large amount of heat is given off during development of tension and during contraction; this is called the *initial heat*. The initial heat can be in turn subdivided into two phases. Some of it is given off just by production of tension, before shortening occurs, whereas the rest appears during the actual shortening process. When the muscle relaxes again after a single contraction, there is still another phase of heat production that lasts much longer than the initial heat production. This is called the *recovery heat*. The recovery heat is a reflection of the processes by which the muscle system is recharged with chemical energy by respiration. Measurements of heat production and of the tension produced also permit the determination of the efficiency of muscle action. This is most simply given by the expression

$$\text{efficiency} = \left(\frac{\text{mechanical work performed}}{\text{work performed} + \text{heat produced}}\right) 100$$

Most skeletal muscles under optimum conditions are about 20 per cent efficient.

11–3 THE ENERGY SOURCE FOR MUSCULAR CONTRACTION

In vertebrates the muscles are supplied with fuels and oxygen by the blood. It is a matter of common experience that muscles use more oxygen when they are active; we tend to puff and blow after the exertion of swimming a hundred yards, for example. However, it has been found that oxygen or the process of respiration is not an absolute requirement for muscular contraction. Muscles can also contract under fully anaerobic circumstances or when poisoned with cynanide. In such anaerobic muscles, where does the energy for contraction come from? Chemical analysis of muscles reveals that the amount of glycogen, the storage form of glucose in muscle, decreases during anaerobic contraction. Concomitant with the decrease in glycogen there is a proportional increase in lactic acid. Muscle thus can perform work at the expense of the energy liberated during the anaerobic breakdown of glucose to lactic acid. However, when muscle contracts anaerobically it uses up much more glucose than it does aerobically. This can be expected because we know that glycolysis delivers only two molecules of ATP from each molecule of glucose, whereas the aerobic oxidation of one molecule of glucose to CO_2 and H_2O delivers thirty-six molecules of ATP (Chapter 5). Moreover, isolated muscles cannot work very long under anaerobic conditions, since under these conditions they use up their glycogen very rapidly. Nevertheless, the ability to contract anaerobically is very useful physiologically because during short bursts of tremendous exertion oxygen cannot be supplied to the muscles fast enough for maximum respiration.

It is also possible to "uncouple" the contractile system of muscle from glycolysis or respiration and cause the muscle to perform mechanical work without concomitant utilization of glucose. An isolated anaerobic muscle will continue to contract even if we prevent the conversion of glucose to lactic acid by applying to the muscle an inhibitor capable of blocking the action of one of the glycolytic enzymes, such as *iodoacetate* (ICH_2COO^-), which inhibits 3-phosphoglyceraldehyde dehydrogenase. Under these conditions no ATP can be generated in an anaerobic muscle. Nevertheless, the iodoacetate-poisoned muscle will contract perfectly well on stimulation, although it cannot give as many contractions as an unpoisoned muscle. This important discovery, which was made by Lundsgaard in 1931, called for an explanation because such muscles appeared to be working with no evident utilization of fuel.

When muscles poisoned with iodoacetate were analyzed, it was found that important chemical changes were indeed taking place during contraction, not in the content of glycogen, fat, or protein but rather in the amount of a specific, high-energy phosphate compound. Unexpectedly, it was not ATP but rather the compound *phosphocreatine*, the phosphorylated derivative of the nitrogenous substance *creatine*:

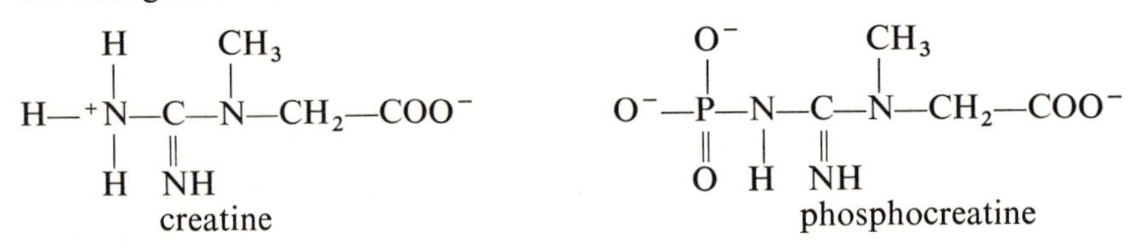

Phosphocreatine rapidly disappears during contraction of iodoacetate-poisoned muscle, with the formation of corresponding amounts of free creatine and inorganic phosphate. During this process, however, there is no decline in the ATP content of muscle. At first this fact appeared rather damaging to the concept that ATP is the direct fuel for the performance of muscular work, since it suggested that phosphocreatine is the immediate fuel for muscular contraction. But an answer to this paradox ultimately appeared following some years of research on the relationship between phosphocreatine and ATP.

11–4 PHOSPHOCREATINE: A HIGH-ENERGY PHOSPHATE RESERVOIR

Phosphocreatine is a high-energy phosphate compound, as was mentioned briefly in Chapter 3. It exists in rather large concentrations in a few tissues of higher animals, particularly in muscle and nerve cells. In skeletal muscle there may be over five times as much phosphocreatine as ATP. Phosphocreatine undergoes only one known enzymatic reaction in muscle

$$\text{phosphocreatine} + \text{ADP} \rightleftharpoons \text{creatine} + \text{ATP} \qquad (11\text{–}1)$$

catalyzed by the enzyme *creatine phosphokinase*. This reaction is reversible. However, the point of equilibrium, if we start from equimolar concentrations of phosphocreatine and ADP, will lie far to the right, as we can predict from the phosphate transfer potential scale in Chapter 3. This scale shows that phosphocreatine has a much more negative standard free energy of hydrolysis than ATP and will therefore tend to donate its phosphate group to ADP under standard conditions. This equilibrium is thus poised in such a manner as to keep essentially all the ADP phosphorylated as ATP at all times, at the expense of phosphocreatine. For this reason it is the phosphocreatine rather than ATP that declines in concentration when iodoacetate-poisoned muscle contracts.

Recently it has been found possible to poison creatine phosphokinase in intact muscles. Under these conditions phosphocreatine does not disappear during contraction, but ATP does, with formation of ADP and phosphate. From such experiments it has been concluded that when muscle contracts, ATP is simultaneously hydrolyzed to ADP plus phosphate, presumably by the enzyme molecules of the contractile machinery. However, as quickly as ADP is formed it is rephosphorylated. If the muscle is aerobic ADP is rephosphorylated during oxidative phosphorylation by the mitochondria. If it is anaerobic, ADP is rephosphorylated at the expense of glycolysis. If respiration or glycolysis fail or are poisoned, ADP is rephosphorylated by phosphocreatine. In effect, the ATP concentration of muscle is always maintained at a high and constant steady-state concentration by these three systems. The phosphocreatine pool represents a labile reservoir of high-energy phosphate groups. It must be regenerated again from free creatine at the expense of ATP formed by oxidative phosphorylation during the recovery period.

Skeletal muscle therefore is biochemically adapted to work on an intermittent basis and has the capacity to function maximally for short periods in the absence of oxygen.

11–5 THE STRUCTURE OF MUSCLE CELLS

From microscopic examination of skeletal muscle cells, important information as to the nature of the molecular machines responsible for converting chemical to mechanical energy has been deduced. Skeletal muscle is made up of parallel bundles of very large multinucleated cells. These muscle cells in turn contain a number of parallel *myofibrils*, which are the basic functional units of the contractile machinery. Muscle cells have a cross-striated appearance due to occurrence of regularly alternating lateral zones in which the optical or refractile properties of the myofibril substances are sharply different. These cross-striations are shown in Fig. 11–1. In very active muscles, such as in the flight muscles of insects, mitochondria are usually found adjacent to the cross-striations in a regular array, so as to juxtapose ATP-producing and ATP-utilizing structures.

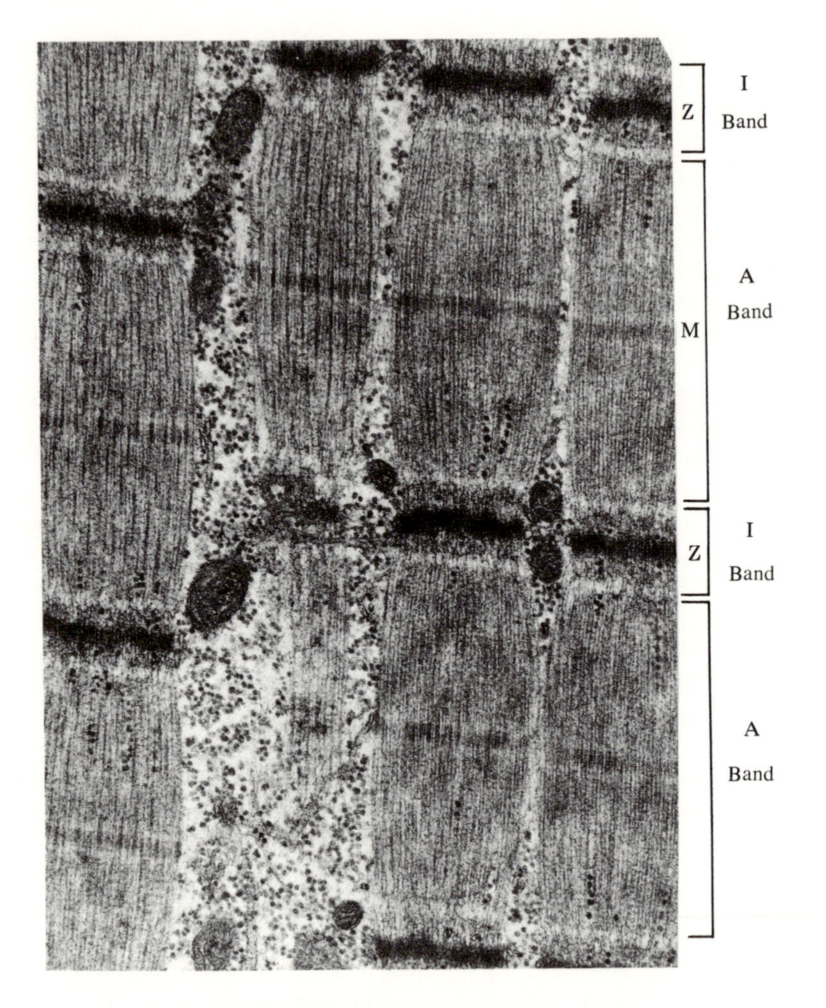

Figure 11–1. Electron micrograph of skeletai muscle, showing dark and light cross-striations (courtesy of Dr. Ronald Bergman, Johns Hopkins University).

Figure 11–2 is a diagrammatic representation at higher magnification of a single myofibril showing the cross-striations and other structures in more detail. The recurring units along the long axis of the myofibril, between two successive Z lines, are called *sarcomeres*. In each sarcomere there is a central M *line*, a central H *disk*, and then two dark zones symmetrically located on either side of the H disk; these striations together constitute the A *band*. The letter A stands for *anisotropic*. This zone possesses *optical anisotropy*; that is to say, its refractive index varies depending on the direction of measurement;

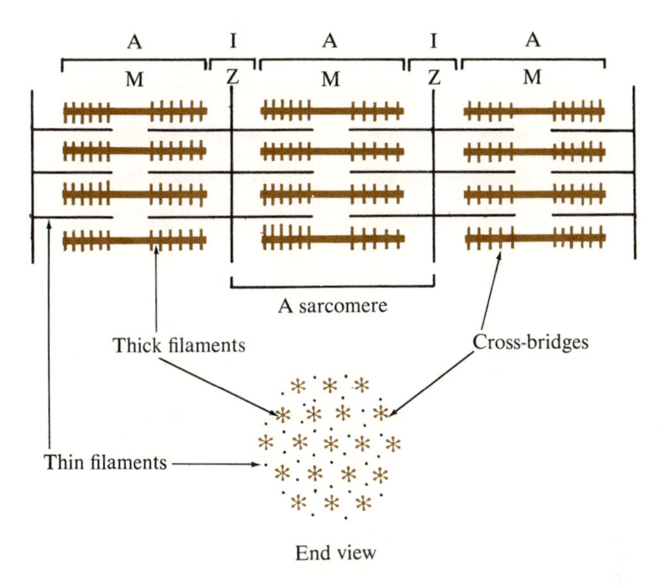

Figure 11–2. Schematic representation of ultrastructure of myofibril, showing sarcomere, the thick and thin myofilaments, and the A and I bands.

this zone is also spoken of as having *double refraction or birefringence*. The light zone between two successive A bands is the I *band* or *isotropic band*; it shows very little birefringence.

The A bands contain many parallel thick filaments in hexagonal array; they run the entire length of the A band. The I bands contain many parallel thin filaments that run into the A bands and end at the edge of the H zone.

The geometrical changes that take place in the filaments when the myofibril contracts (Fig. 11–3) have been deduced from ingeniously designed electron microscope and optical measurements. When the myofibril shortens during contraction, the A band remains fixed in length but the I band shortens and may completely disappear in the fully contracted muscle; the H disk may also disappear. On relaxation of the myofibril the original configuration is restored. From such findings it has been postulated that the sets of thick and thin filaments that are present in the A and I bands, respectively, slide along each other during contraction so that the thin filaments move into the A bands, whose heavy filaments remain fixed in position.

11–6 MYOSIN, ACTIN, AND ACTOMYOSIN

Two major protein components have been discovered in extracts of muscles, which are now known to constitute the structural elements of the two types of

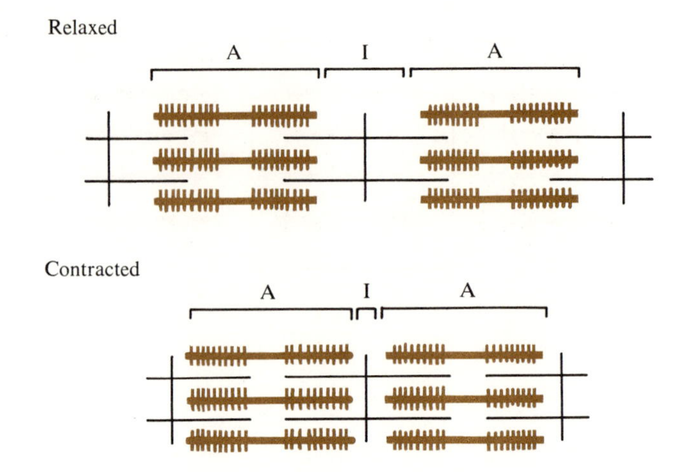

Figure 11–3. The sliding filament model of muscle contraction. The thin filaments slide between the fixed thick filaments to cause shortening of the I bond.

filaments in the myofibril. The first to be found is the protein *myosin*, which makes up about 50–55 per cent of the myofibrillar protein of skeletal muscle in the mammal. This protein was recognized as a component of the contractile system by virtue of the fact that when it is extracted from muscle with strong salt solutions, the remaining extracted muscle no longer shows the A or anisotropic bands. Morever, the isolated myosin protein itself shows double refraction when the molecules are oriented in one direction. From these observations it has been concluded that myosin must be the substance of which the heavy filaments of the A zone consist.

Myosin is a long, thin, highly charged protein molecule having a unit molecular weight of about 470,000. Its structural organization is given in Fig. 11–4. It consists of two long helical polypeptide chains, each having about 1800 amino acid residues; the chains are supercoiled around each other. The molecule has a head at one end, which also contains two short polypeptide chains. Myosin therefore has a complex molecular architecture, but its most spectacular property lies in the fact it has enzymatic activity. It catalyzes the hydrolysis of ATP to ADP and phosphate in the presence of Ca^{2+}. Moreover, it has been found that the ATP-hydrolyzing activity is localized entirely in the head of the myosin molecule; the tail of myosin can be split off without loss of catalytic activity.

The second major protein component associated with the contractile system can be extracted from muscle by a somewhat different procedure than that used to obtain myosin. This protein has been called *actin*, and it makes up 20–25 per cent of the fibrillar protein. Actin has been isolated in two forms.

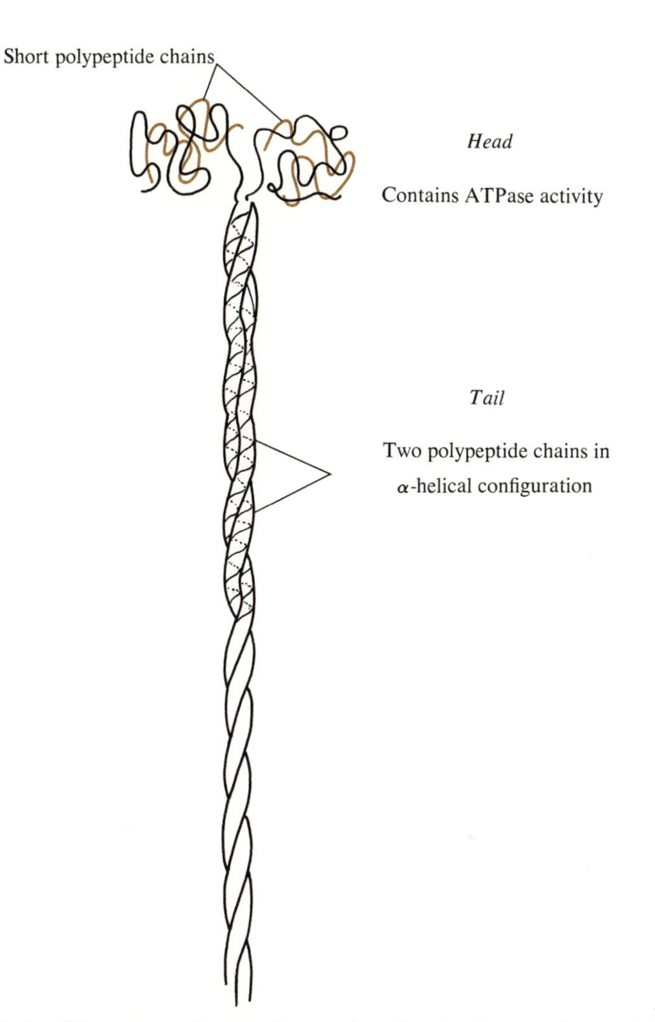

Figure 11–4. Diagram of myosin molecule (redrawn from Albert L. Lehninger, *Biochemistry*, Worth Publishers, New York, 1970, p. 589).

One form is a monomeric globular protein with a molecular weight of about 46,000; it is called *globular actin* or *G-actin.* The other is a long, stringy two-stranded structure made up of recurring G-actin units polymerized in a linear fashion. This is called *fibrous actin* or *F-actin.* When G-actin is mixed with ATP in the test tube, the ATP is broken down to ADP and phosphate, and the G-actin undergoes polymerization to the F-actin form (Fig. 11–5). It now appears probable that the thin filaments of the I bands in the myofibril consist of strands of F-actin.

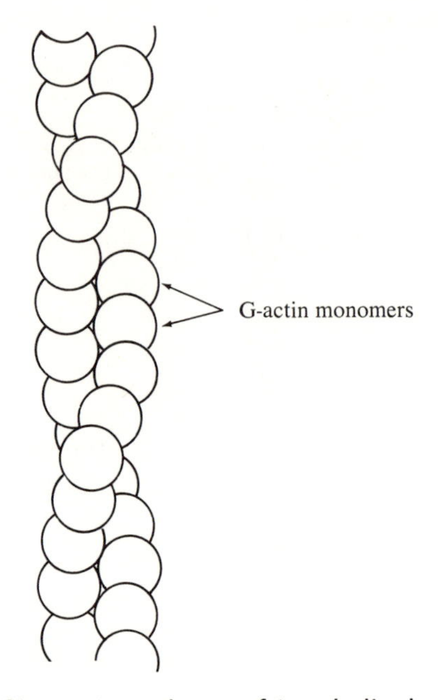

Figure 11–5. F-actin filament, made up of two helical chains of G-actin monomers.

When the conditions for extraction of muscle tissue are properly adjusted, a complex of actin and myosin called *actomyosin* can be extracted. Actomyosin can also be made in the test tube, by mixing purified actin and myosin. A solution of actomyosin has very high viscosity, indicating that the complex has a long, rodlike shape. Actomyosin, like actin and myosin, also has the capacity to hydrolyze ATP to ADP and phosphate, and its activity in this respect is much greater than that of myosin itself. Moreover, Ca^{2+} ions are required for ATP hydrolysis by actomyosin. During the hydrolysis of ATP the viscosity of the solution falls sharply, but when the ATP has been completely hydrolyzed, the viscosity returns to its original high value. These viscosity changes have led to the conclusion that the hydrolysis of ATP by actomyosin is accompanied by dissociation of the latter to form free actin and free myosin; when the ATP is exhausted the components recombine again to form actomyosin. That actomyosin undergoes a physical change in the presence of ATP can be demonstrated in another way. If a solution of actomyosin is rapidly extruded from a hypodermic syringe into water, the actomyosin precipitates as a thread or fiber. If ATP is now added to such an artificial thread of actomyosin, it contracts and develops a weak tension.

11–7 THE MOLECULAR MECHANISM OF MUSCULAR CONTRACTION

Important clues to the mechanism and geometry of the interaction between F-actin filaments and myosin have come from both electron microscopy and biochemical studies of muscle. High-resolution electron microscopy has revealed that the thick heavy filaments of the A bands in intact muscle contain many barblike projections regularly arranged in a helical fashion around the filament. These projections extend to and appear to make contact with the six adjacent thin filaments. The barblike projections have been identified as the heads of individual myosin molecules. From such studies it has been deduced that the thick filaments consist of parallel bundles of myosin molecules, with the latter so arranged that their heads project outwards at regular intervals.

Biochemical studies have shown that when actin and myosin combine to form the actomyosin complex it is the projecting head of the myosin molecule that combines with sites on the F-actin filament. Since the head of the myosin molecule has the ATPase activity and since actin and myosin dissociate in the presence of ATP, but recombine when ATP is hydrolyzed to ADP, it has been concluded that ADP favors the making of lateral cross-bridges between the thick and thin filaments, by allowing the interaction of the myosin heads with the actin filament, whereas ATP favors the breaking or loosening of these cross-bridges.

From this picture we may now visualize how the thin filaments are made to slide along the thick filaments at the expense of ATP energy (Fig. 11–6). It has been suggested that as each cross-bridge is loosened, a process that requires hydrolysis of one molecule of ATP, that portion of the myosin molecule near the head may undergo a hingelike bending or perhaps a shortening, which has the effect of displacing the head along the actin filament. After the hydrolysis of the ATP in this local area is complete, the myosin head now forms a new cross-bridge, further down the actin filament. In this hypothesis, which is supported by much evidence, the barblike heads of the myosin molecules in the thick filaments pull (or push) the thin filaments along toward the center of the A band, causing contraction, through the successive making and breaking of the many cross-bridges. Each time a cross-bridge is broken, a translocation of a small segment of the thin filament is brought about.

11–8 RELAXATION OF MUSCLE

Since ATP is present in high concentrations in resting muscle and is constantly reformed from ADP by glycolysis and oxidative phosphorylation, we may well ask why muscle fibers do not remain contracted at all times. What are the factors that actually initiate contraction? How does the muscle fiber relax again?

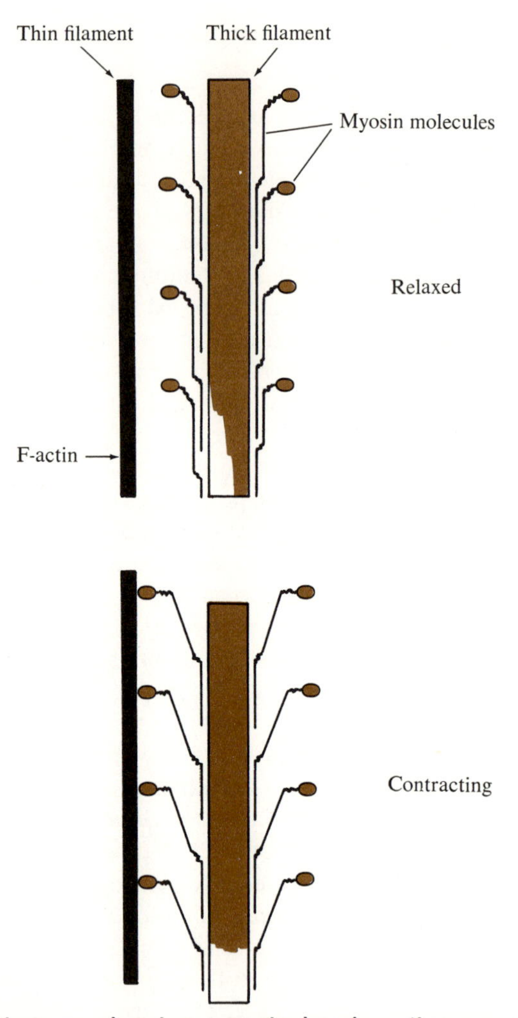

Figure 11–6. Diagram showing myosin heads acting as cross-bridges during contraction (redrawn from Albert L. Lehninger, *Biochemistry*, Worth Publishers, New York, 1970, p. 593).

The answers to these questions can be found in the important role of Ca^{2+} ions in muscle. We have noted above that Ca^{2+} is required for the ATPase activity of the actomyosin system; little or no ATP hydrolysis occurs in its absence. The presence of free ionic Ca^{2+} thus would promote the breaking of the cross-bridges between the thick and thin filaments of muscle and the subsequent shortening of the sarcomere, whereas the absence of free Ca^{2+} would favor maintenance of the cross-bridges in the relaxed form. But how is free ionic Ca^{2+} introduced into the neighborhood of the thick and thin filaments to initiate contraction? How is Ca^{2+} again removed to relax the muscle?

When the motor nerve impulse arrives at the junction of the nerve with the surrounding membrane of the muscle, it excites not only the plasma membrane of the muscle cell, which is called the *sarcolemma*, but also a complex system of internal tubular membranes that connect with the sarcolemma and run across the muscle cell at regular intervals. These are called *sarcotubules* and the entire system of transverse tubules is called the *T system*; it also connects with the endoplasmic reticulum. The excitation of this complex membrane system causes its membrane to depolarize, that is, to lose the electrical potential that normally exists across it. This is caused by a momentary increase in its permeability. As a consequence, some of the internal electrolytes within the tubule escape into the sarcoplasm, among which are Ca^{2+} ions, which are normally segregated within the tubules of resting muscle. The Ca^{2+} ions so released thus come into contact with the actin and myosin filaments, which are surrounded by the ATP-rich sarcoplasm. The arrival of the Ca^{2+} immediately causes the breaking of the cross-bridges between the actin and myosin filaments and, concomitantly, the translocation of the actin filaments and the hydrolysis of the ATP to ADP and phosphate. As ADP accumulates locally, the cross-bridges reform; many cycles of rapid making and breaking of the cross-bridges occur with continuing contraction so long as free Ca^{2+} is available in the sarcoplasm.

Relaxation of the muscle requires that the free Ca^{2+} in the sarcoplasm be removed again, by its return to the internal lumen of the sarcotubules. After the T system recovers from the excitation process and regains its normal resting impermeability to Ca^{2+}, the Ca^{2+} ions in the sarcoplasm are "pumped" back into the tubules by means of an ATP-dependent active transport system in the membrane of the tubules. This system resembles the Na^{+}-, K^{+}-stimulated ATPase on the erythrocyte membrane, in that Ca^{2+} is pumped across the membrane in a directional manner at the expense of the energy released by ATP hydrolysis. For every ATP molecule hydrolyzed, two Ca^{2+} ions are pumped back into the tubule. This pump continues to function and utilize ATP until the concentration of Ca^{2+} in the sarcoplasm reaches the very low level of 1×10^{-7} M, at which point the fiber is relaxed; ATP utilization then stops. So long as Ca^{2+} remains segregated in the tubules and sarcoplasmic reticulum, the muscle remains relaxed. However, when the next motor nerve impulse arrives, this Ca^{2+} is immediately released into the sarcoplasm again at a very high rate, which enables all the filaments to respond simultaneously.

We see, therefore, that ATP plays three roles in muscular contraction. It is required for the breaking of the cross-linkages between the actin and myosin filaments, for the translocation process, and for the active transport of Ca^{2+} from the sarcoplasm back into the T system, to promote relaxation.

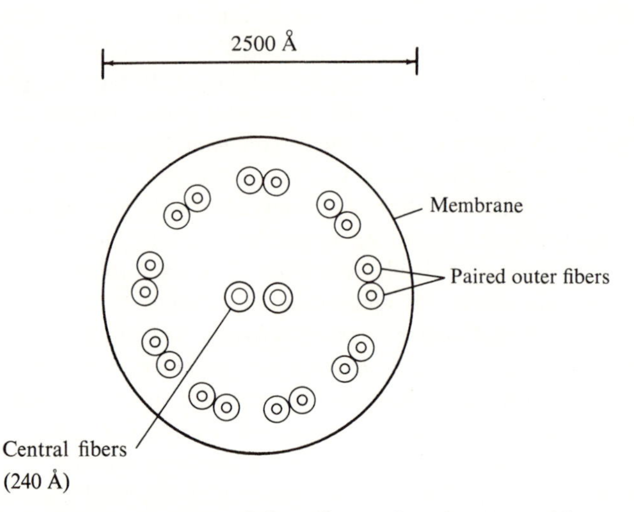

Figure 11–7. Cross section of flagellum showing "9 + 2" arrangement of paired tubular fibers.

11–9 FLAGELLA AND CILIA

The characteristic gyratory or sweeping motion of the flagella and cilia of eukaryotic cells is "powered" by the hydrolysis of ATP. The cilia and flagella of higher organisms consist of a group of eleven fibers in a matrix surrounded by a membrane. The eleven fibers are arranged in a characteristic 9 + 2 pattern (Fig. 11–7). Each of the eleven fibers consists of a pair of microtubules, each of which in turn is made up of twelve filaments of small globular units. The two inner fibers are larger in diameter than the nine outer fibers. At the base of the cilium or flagellum is a large organized structure, about the size of a mitochondrion, called the *kinetosome*, which apparently furnishes the ATP.

Cilia and flagella have been isolated following their detachement from cells. When ATP and Mg^{2+} are added to suspensions of isolated flagella, they undergo wavelike oscillations; simultaneously the ATP undergoes hydrolysis. The ATPase activity appears to be localized in the outer fibers.

The precise manner in which ATP causes flagella to undergo oscillations is not known with certainty, but it appears likely that the paired outer fibers undergo cycles of twisting and untwisting around each other.

12

SOME OTHER PROBLEMS IN BIOENERGETICS

In this book we have traced the energy transformations that occur in the mainstreams of energy flow in living organisms, those involved in growth, maintenance, replication, and locomotion of cells. However, some of the most fascinating problems in bioenergetics are posed by processes in which relatively minor and sometimes only minute amounts of energy are exchanged. Often these are very important for the survival of the organism. Let us briefly sketch a few of these energy transformations in order to show the challenging questions they raise for the molecular biologist. Some of them are concerned with the perception of energy changes in the environment and others with the emission of energy for communication or navigation.

Perhaps the most conspicuous energy-sensing devices in higher animals are the light-sensitive cells of the retina of vertebrates. The human eye is an extremely accurate light receptor that has a high degree of discrimination, not only for different intensities of light but also for different colors or wavelengths. Moreover, the eye has the ability to resolve or distinguish between closely juxtaposed points in the visual arc. The light-sensitive rod and cone cells in the retina contain light-absorbing pigments. One of these is called *rhodopsin*; it is a complex of a protein called *opsin* and a derivative of vitamin A called 11-*cis*-retinal (Fig. 12–1), which is chemically similar to the light-absorbing carotenoids found in some photosynthetic organisms. When light

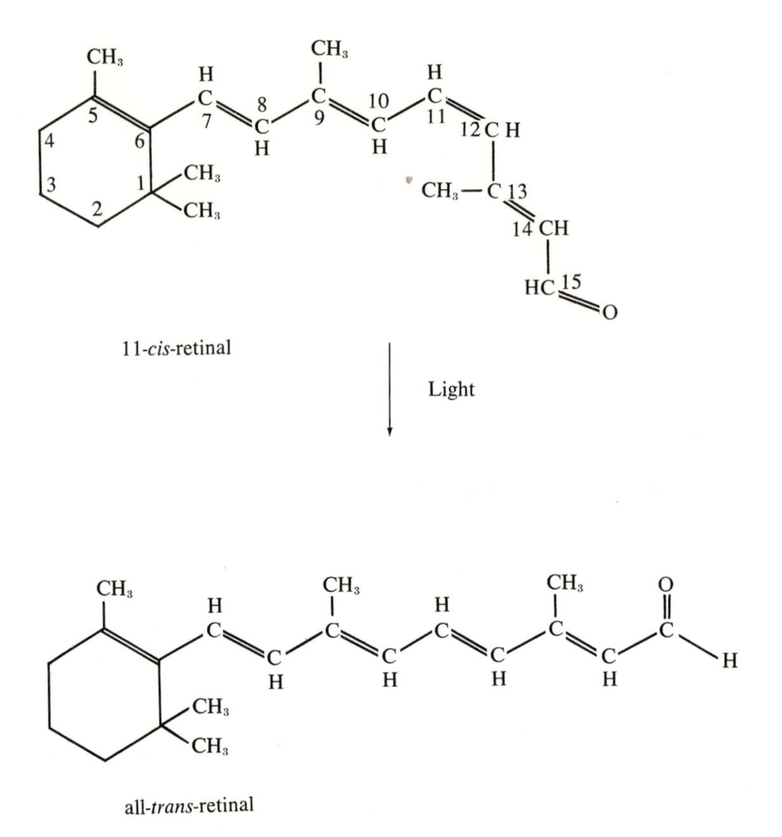

Figure 12–1. The light-dependent configurational change in 11-*cis*-retinal of retinal rods.

energy is absorbed by rhodopsin the 11-*cis*-retinal undergoes a configurational change, a *cis*→*trans* isomerization, to yield all-*trans*-retinal. This configurational change is in some as yet unknown way transmitted to the membranes of the rod cells, which in turn excite certain nerve cells leading to the brain. After this excitation process, the all-*trans*-retinal is converted back to the 11-*cis*-retinal by an enzyme, thus completing what is called the visual cycle of the retinal rod.

Some organisms have evolved sensory systems that can detect extremely small changes in the temperature of the surroundings. This capacity is developed to an extraordinary degree in such animals as the pit vipers. The pits of these snakes, once thought to be ears, contain cells that function as temperature sensors. These structures can detect the presence of nearby prey or a predator through the minute amounts of heat energy they emanate. The physical or chemical basis by which such heat-sensing cells act is not known, but presumably the absorption of heat energy brings about a conformational change of specific temperature-sensitive macromolecules in these cells. Since protein molecules respond to temperature increases (or decreases) by changes in the configuration of their polypeptide chains, it appears likely that specific proteins in these cells have been adapted to act as exquisitely sensitive heat detectors.

The ability to perceive electrical energy is also extremely highly developed in some organisms. Some fishes can detect minute changes in electrical energy in their environment. Among these is one that, when swimming in an aquarium, can detect the static electricity emanating as a nearby person combs his hair. The sensitivity of these receptors may exceed the sensitivity of man-made devices for measurement of electrical energy. The molecular mechanisms by which electrical energy is absorbed by specialized cells in the midline of these fishes and transformed into a nerve impulse are still unknown.

Some organisms are capable of emitting energy in one form or another. The bat can emit high-frequency sonic signals, which it uses in echo location and navigation. Some fishes, such as those mentioned above, can emit electrical signals that are also involved in navigation or hunting. Sometimes emission of energy is used for protective purposes. For example, the electric eel can deliver stunning shocks of several hundred volts. Its electric organ is the subject of intensive biochemical, physiological, and ultrastructural study. The bombardier beetle has a miniature "cannon" that fires a tiny puff of smoke, through the pressure–volume energy developed as hydrogen peroxide is enzymatically decomposed into gaseous oxygen and water.

Still another type of biological energy emission is *bioluminescence*, the conversion of chemical energy into light energy. This conversion is most familiar and most highly developed in the mating signal of the firefly, which contains in its lantern an enzyme (*luciferase*) capable of utilizing the energy of ATP to induce the excited state of an organic substance called *luciferin* (Fig. 12–2). In the luciferin molecule certain electrons are raised from low-energy to high-energy orbitals in the presence of ATP. It is the

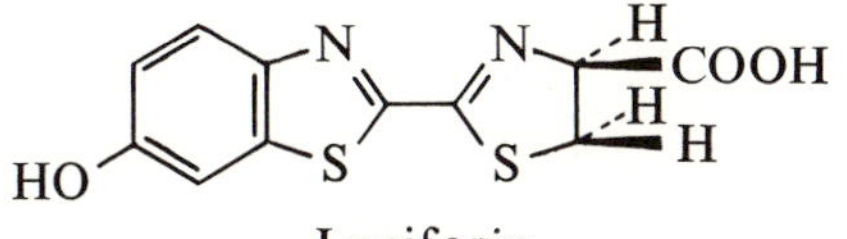

Luciferin

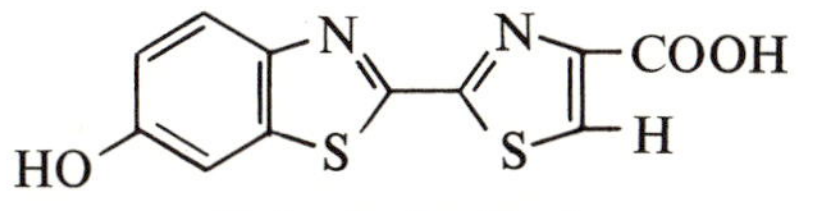

Dehydroluciferin

Figure 12–2. The structure of luciferin and its oxidation product.

energy-yielding return of these high-energy electrons to the ground state, as luciferin is oxidized to dehydroluciferin, which yields the luminescence, according to the overall equation

$$\text{luciferin} + \text{ATP} + \tfrac{1}{2}\text{O}_2 \longrightarrow \text{dehydroluciferin} + \text{AMP} + \text{PP}_i + \text{light}$$

Perhaps the most challenging of all problems in biogenergetics is posed by the phenomenon of memory, particularly the highly developed ability of man to store relatively immense amounts of information in his memory. We have already seen that information has a thermodynamic dimension, since it is the opposite of entropy. Thus there must be a finite energy cost wherever information is laid down in memory. How memory is stored in the human brain, whether in the form of specific interneuronal connections or in the form of macromolecular structure, such as is used for storage of genetic information, is still totally unknown. Perhaps the molecular basis of memory, thought, and behavior will be among the ultimate biological secrets to be solved by man, secrets that will come close to defining man's own being. Thus we see that in its broadest aspects bioenergetics spans some of the most remarkable and specialized activities in the world of living organisms.

REFERENCES

General textbooks

Dickerson, R. E., Gray, H. B., and Haight, G. P., Jr., *Chemical Principles*, Benjamin, New York, 1970.

DuPraw, E. J., *Cell and Molecular Biology*, Academic, New York, 1968.

Keeton, W. T., *Biological Science*, Norton, New York, 1967.

Lehninger, A. L., *Biochemistry: The Molecular Basis of Cell Structure and Function*, Worth, New York, 1970.

Loewy, A. and Siekevitz, P., *Cell Structure and Function*, 2nd ed., Holt, New York, 1969.

Moore, W. J., *Physical Chemistry*, 3rd ed., Prentice-Hall, Englewood Cliffs, New Jersey, 1962.

Morrison, R. T. and Boyd, R. N., *Organic Chemistry*, 2nd ed., Allyn and Bacon, Inc., Boston, 1966.

Raven, P. H. and Curtis, H., *Biology of Plants*, Worth, New York, 1970.

Stanier, R. Y., Doudoroff, M. and Adelberg, E. A., *The Microbial World*, 3rd ed., Prentice-Hall, Englewood Cliffs, N. J., 1970.

Watson, J. D., *Molecular Biology of the Gene*, 2nd ed., Benjamin, New York, 1970.

Chapter 1

Grobstein, C., *The Strategy of Life*, Freeman, San Francisco, 1965.

Loewy, A. and Siekevitz, P., *Cell Structure and Function*, 2nd ed., Holt, New York, 1969.

Chapter 2

Blum, H. F., *Time's Arrow and Evolution*, Harper, New York, 1962.

Katchalsky, A. and Curran, P. F., *Non-equilibrium Thermodynamics in Biophysics*, Harvard, Cambridge, 1965.

Klotz, I. M., *Energy Changes in Biochemical Reactions*, Academic, New York, 1967.

Morowitz, H. J., *Energy Flow in Biology*, Academic, New York, 1968.

Morowitz, H. J., *Entropy for Biologists*, Academic, New York, 1970.

Chapter 3

Ingraham, L. L. and Pardee, A. B., "Free Energy and Entropy in Metabolism," in D. M. Greenberg (ed.) *Metabolic Pathways*, 3rd ed., Vol. 1., Academic, New York, 1967, p. 2.

Kalckar, H. M., *Biological Phosphorylations*, (A collection of classical papers.) Prentice-Hall, Englewood Cliffs, N. J., 1969.

Kaplan, N. O. and Kennedy, E. P. (eds.), *Current Aspects of Bioenergetics*, (A volume of essays dedicated to Fritz Lipmann.) Academic, New York, 1966.

Krebs, H. A. and Kornberg, H. L., *Energy Transformations in Living Matter*, (In English) Springer, Berlin, 1957.

Chapter 4

Axelrod, B., "Glycolysis," in D. M. Greenberg (ed.), *Metabolic Pathways*, 3rd ed., Vol. 1., Academic, New York, 1967, p. 112.

Bernhard, S., *Enzymes: Structure and Function*, Benjamin, New York, 1968.

Koshland, D. E., Jr. and Neet, K. E., "The Catalytic and Regulatory Properties of Enzymes," *Annual Reviews of Biochemistry*, 37, 359 (1968).

Stadtman, E. R., "Allosteric Regulation of Enzyme Activity," *Adv. in Enzymology*, 28, 41 (1966).

Chapter 5

Florkin, M. and Stotz, E. H. (eds.) "Biological Oxidations," in *Comprehensive Biochemistry*, Vol. 14, Elsevier, New York, 1967.

Keilin, D., *The History of Cell Respiration and Cytochromes*, Cambridge Univ. Press, New York, 1966.

Lehninger, A. L., *The Mitochondrion: Molecular Basis of Structure and Function*, Benjamin, New York, 1964.

Lowenstein, J. M., "The Tricarboxylic Acid Cycle," in D. M. Greenberg (ed.), *Metabolic Pathways*, 3rd ed., Vol. 1, Academic, New York, 1967, p. 146.

Mitchell, P., "Chemiosmotic Coupling in Oxidative and Photosynthetic Phosphorylation," *Biol. Rev.* 41, 445 (1965).

Racker, E., ed. *Structures and Function of Membranes of Mitochondria and Chloroplasts*, Reinhold, New York, 1969.

Chapter 6

Arnon, D., "Photosynthetic Activity of Isolated Chloroplasts," *Physiol. Rev.* 47, 317 (1967).

Boardman, N. K., "The photochemical system of photosynthesis," *Adv. in Enzymology 30*, 1 (1968).

Calvin, M. and Bassham, J. A., *The Photosynthesis of Carbon Compounds*, Benjamin, New York, 1962.

Hill, R., "The Biochemist's Green Mansions: The Photosynthetic Electron Transport Chain in Plants" in P. H. Campbell and G. D. Greville (eds.) *Essays in Biochemistry*, 1, 121, Academic, New York, 1965.

Moudrianakis, E., "Structural and Functional Aspects of Photosynthetic Lamellae, (A symposium article.) *Federation Proc.*, 27, 1180 (1968).

Chapter 7

Dagley, S. and Nicholson, D. E., *An Introduction to Metabolic Pathways*, Wiley, New York, 1970.

Greenberg, D. M., ed., *Metabolic Pathways*, Vol. 1-3, Academic, New York, 1967–1969.

Ingram, V., *Biosynthesis of Macromolecules*, 2nd ed., Benjamin, New York, 1970.

Stadtman, E., "Allosteric Regulation of Enzyme Activity," *Adv. in Enzymology 28*, 41 (1966).

Chapter 8

Cold Spring Harbor Symposia in Quantitative Biology

Vol. 31, *The Genetic Code* (1966).

Vol. 33, *Replication of DNA in Microorganisms* (1968).

Vol. 34, *Mechanism of Protein Synthesis* (1969).

Vol. 35, *Transcription of Genetic Material* (1970).

Watson, J. D., *Molecular Biology of the Gene*, 2nd ed., Benjamin, New York, 1970.

Ycas, M., *The Biological Code*, North-Holland, Amsterdam, 1969.

Chapter 9

Dickerson, R. E. and Geis, I., *The Structure and Action of Proteins*, Harper and Row, New York, 1969.

Fraenkel-Conrat, H., *Design and Function at the Threshold of Life: The Viruses*, Academic, New York, 1962.

Lehninger, A. L., *Biochemistry*, Chapter 33. Worth, New York, 1970.

Nomura, M., "Ribosomes," *Scientific American*, December, 1969.

Wood, W. D. and Edgar, R. S., "Building a bacterial virus," in R. H. Haynes and P. C. Hanawalt, (eds.) *Molecular Basis of Life*, Freeman, San Francisco, 1968.

Chapter 10

Mitchell, P., "Translocations Through Natural Membranes," *Adv. in Enzymology 29*, 33 (1968).

Stein, W. D., *Movement of Molecules Across Cell Membranes*, Academic, New York, 1967.

Chapter 11

Davies, R. E., "On the Mechanism of Muscular Contraction" in P. N. Campbell and G. D. Greville, *Essays in Biochemistry*, Vol. 1, Academic, New York, 1965, p. 29.

Gibbons, I. R., "The Biochemistry of Motility," *Ann. Rev. Biochem.*, 37, 521 (1968).

Huxley, H. E., "The Mechanism of Muscular Contraction," *Science*, 164, 1356 (1969).

Smith, D. S., "The Flight Muscles of Insects," *Scientific American*, 212, 76 (1965).

GLOSSARY OF FREQUENTLY USED TERMS

Active site Region of enzyme surface which interacts with the substrate molecule.

Actomyosin A molecular complex of two muscle proteins, actin and myosin; the basic contractile element in muscle.

Aerobes Cells that live in and utilize oxygen; strict aerobes cannot live without oxygen.

Allosteric enzymes Enzymes whose activity is modulated by the binding of specific small metabolites (allosteric effectors) at sites other than the active site. Also called regulatory enzymes.

Aminoacyl synthetases Enzymes that catalyze the ATP-dependent reaction of a specific amino acid with its tRNA to yield an aminoacyl -tRNA, AMP, and pyrophosphate.

Anaerobes Cells that can live without oxygen; strict anaerobes cannot live in the presence of oxygen.

Angstrom (Å) A unit of length used to indicate atomic dimensions. It is equal to 10^{-8} cm.

Autotrophic cells Cells that can build their own macromolecules from very simple nutrient molecules, such as carbon dioxide and ammonia.

Bit A binary digit, the basic unit of information content. It represents the answer to a binary choice, i.e., two mutually exclusive possibilities such as "yes" and "no."

Bond energy The energy required to break a bond. It must be differentiated from the term "phosphate bond energy" as used in biology.

Calorie A measure of energy; the amount of heat required to raise the temperature of 1.0 gm of water from 14.5° to 15.5°C.

Central dogma The principle that genetic information flows from DNA to RNA to protein.

Chloroplasts Membrane-surrounded structures containing chlorophyll which are present in the cytoplasm of eukaryotic photosynthetic cells; they are the sites of conversion of light energy into chemical energy.

Closed system A system which does not exchange matter with the surroundings, although it may exchange energy.

Codon A sequence of three adjacent nucleotides in a nucleic acid that codes for an amino acid.

Common intermediate A chemical compound that is common to two chemical reactions, as either a reactant or a product. Chemical energy may be transferred from one reaction to another via such a common intermediate.

Coupled reactions Two chemical reactions which have a common intermediate and therefore a means by which energy can be transferred from one reaction to the other.

Dalton The weight of a single hydrogen atom. (1.67×10^{-24}g).

Diffusion The tendency for molecules to move in the direction of a lesser concentration, so as to make the concentration uniform throughout the system and the entropy maximum.

Electron carriers Enzymes such as flavoproteins and cytochromes which can gain and lose electrons reversibly; the respiratory chain consists of a series of electron carriers.

Electron transport The movement of electrons from substrates to oxygen catalyzed by the respiratory chain during respiration.

Endergonic reaction A chemical reaction with a positive standard free energy change; an "uphill" reaction.

Endoplasmic reticulum An extensive system of double membranes in the cytoplasm, often coated with ribosomes.

Endothermic process A process in which heat is absorbed.

Entropy The randomness or disorder of a system.

Equilibrium The state of a system in which there are no unbalanced forces still operating and in which its free energy is at a minimum and the entropy of the universe at a maximum.

Equilibrium constant A physical constant designating the specific concentrations or activities of all components of a chemical system at equilibrium at a given temperature.

Eucaryotic cells Cells having nuclear membranes and membrane-surrounded organelles, which divide by mitosis.

Excited state That energy-rich state of an atom or molecule existing after an electron has been moved from its normal stable orbital to an outer orbital having a higher energy level, as a result of the absorption of light energy.

Exergonic reaction A reaction with a negative standard free energy change; a "downhill" reaction.

Exothermic process A process in which heat is evolved.

Facultative cells Cells that can live either in the presence or absence of oxygen.

Feedback (end-product) inhibition Inhibition of the first enzyme in a metabolic pathway by the end-product of that pathway.

Fermentation The energy-yielding enzymatic breakdown of fuel molecules that takes place in certain cells under anaerobic conditions; oxygen is not required in this process.

First law of thermodynamics In all processes the total energy of the universe remains constant.

Free energy That component of the total energy of a system which can do work under conditions of constant temperature and pressure.

Genetic information The information contained in a sequence of nucleotide bases in a DNA or RNA molecule.

Glycolysis That form of fermentation in which glucose is broken down via pyruvic acid to two molecules of lactic acid.

Heterotrophic cells Cells that require complex nutrient molecules such as glucose, amino acids, etc., from which to obtain energy and to build their own macromolecules.

High-energy bond A bond that yields a large (at least 5 kcal/mole) decrease in free energy upon hydrolysis under standard conditions.

High-energy phosphate compound A phosphorylated compound having a highly negative standard free energy of hydrolysis.

Hormone A chemical substance synthesized in one organ that modulates biochemical functions in the cells of another tissue or organ.

Hydrogen bond A weak attractive force, electrostatic in nature, between one electronegative atom and a hydrogen atom covalently linked to a second electronegative atom.

Hydrolysis The cleavage of a molecule into two or more smaller molecules by the addition of a water molecule.

Hydrophilic "Water-loving"; refers to molecules or groups that associate with H_2O.

Hydrophobic "Water-hating"; refers to molecules or groups that are only poorly soluble in water.

Hydrophobic bond The association of nonpolar groups with each other in aqueous solution because of the tendency of the surrounding water molecules to seek their most stable configuration.

Intermediary metabolism The chemical reactions in a cell that transform food molecules into molecules needed as a source of energy and as precursors for cell growth.

In vitro (Latin: in glass). Refers to experiments done on isolated cells, tissues, or cell-free extracts in (glass) reacting vessels.

In vivo (Latin: in life). Refers to experiments done on intact living organisms.

Irreversible process A process in which the entropy of the universe increases.

Isothermal process A process occurring at constant temperature.

Low-energy phosphate compound A phosphorylated compound having a relatively low standard free energy of hydrolysis.

Macromolecules Molecules having molecular weights in the range of a few thousand to hundreds of millions.

Mitochondria Membrane-surrounded organelles in the cytoplasm of aerobic cells which contain respiratory enzyme systems.

Molecular weight The sum of the atomic weights of the constituent atoms in a molecule.

Mutation An inheritable change in a chromosome.

NAD, NADP Nicotinamide adenine dinucleotide and nicotinamide adenine dinucleotide phosphate, carriers of electrons in many enzymatic oxidation-reduction reactions.

Nucleolus Round, granular structure found in nucleus of eucaryotic cells. Involved in rRNA synthesis and ribosome formation.

Open system A system which exchanges matter as well as energy with its surroundings.

Organelles Membrane-surrounded structures found in eucaryotic cell; they contain enzymes for specialized cell functions. Some organelles, including mitochondria and chloroplasts, have DNA and can replicate autonomously.

Oxidation The loss of electrons from a compound; an oxidizing agent is an electron acceptor.

Oxidative phosphorylation The enzymatic phosphorylation of ADP to ATP which is coupled to electron transport along the respiratory chain to oxygen.

Peptide bond A covalent bond between two amino acids in which the α-amino group of one amino acid is bonded to the α-carboxyl group of the other with the elimination of H_2O.

Phosphate bond energy A term used to denote the decline in free energy as one mole of a phosphorylated compound undergoes hydrolysis to equilibrium at pH 7.0 and 25°, in a 1.0 M solution.

Phosphodiester Any molecule that contains two alcohols ROH and R^1OH esterified to phosphoric acid:

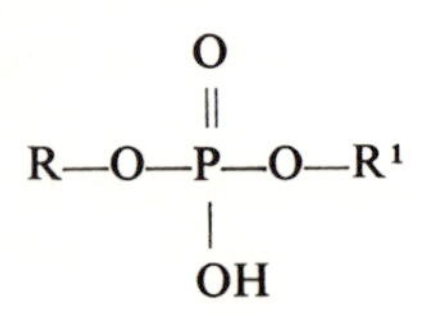

Photosynthetic phosphorylation The enzymatic formation of ATP from ADP in green plants coupled to light-dependent transport of electrons from excited chlorophyll.

Photosynthesis The enzymatic conversion of light energy into chemical energy and use of the latter to form carbohydrates and oxygen from CO_2 and H_2O in green plant cells.

Polynucleotide A linear sequence of nucleotides in which the 3′ position of the sugar of one nucleotide is linked through a phosphate group to the 5′ position on the sugar on the adjacent nucleotide.

Polyribosome Complex of a messenger-RNA molecule and two or more ribosomes.

Primary protein structure The covalent backbone structure of a protein, including its amino acid sequence and the location of inter- and intra-chain disulfide bridges.

Procaryotes Simple unicellular organisms, such as bacteria or blue-green algae, with no nuclear membrane, no membrane-bound organelles, characteristic ribosomes, and a single chromosome.

Puromycin Antibiotic that inhibits polypeptide synthesis by competing with aminoacyl tRNAs for incorporation into the polypeptide chain.

Radioactive isotope An isotope with an unstable nucleus that stabilizes itself by emitting ionizing radiation.

Reduction The gain of electrons by a compound; a reducing agent is an electron donor.

Respiration The oxidative breakdown and release of energy from fuel molecules by reaction with oxygen in aerobic cells.

Reversible process A process which proceeds with no change in entropy.

Ribosomal RNA (rRNA) The nucleic acid component of ribosomes, making up two-thirds of the mass of the ribosome in E. coli; rRNA accounts for approximately 80 percent of the RNA content of the bacterial cell.

Ribosomes Small cellular particles (~ 200 A in diameter) made up of rRNA and protein. Ribosomes are the site of protein synthesis; in eucaryotic cells they are often attached to the endoplasmic reticulum.

RNA (Ribonucleic acid) A polymer of ribonucleotides.

RNA Polymerase An enzyme that catalyzes the formation of RNA from ribonucleoside triphosphates, using DNA as a template.

Second law of thermodynamics The entropy of the universe tends to increase.

Spontaneous process A process accompanied by a decline of free energy.

Standard free energy change The gain or loss of free energy in calories as one mole of reactants in the standard state is converted into one mole of products, also in the standard state, at the equilibrium point of the reaction.

Standard reduction potential The electromotive force exhibited at an electrode by a 1:1 ratio of a reducing agent and its oxidized form at 1.0 at 25°. It is a measure of the relative tendency for the reducing agent to lose electrons.

Standard state The most stable form of a pure substance at 1.0 atmosphere pressure and 25°C (298°K). For reactions occurring in solution, the standard state of a solute is a 1.0 molal solution.

Steady state A nonequilibrium state of an open system through which matter is flowing and in which all components remain in constant concentration.

Substrate The specific compound acted upon by an enzyme.

System An isolated collection of matter. All other matter in the universe apart from the system is called the outside world or surroundings.

Template The macromolecular mold or pattern for the synthesis of another macromolecule.

Tertiary struction (of a protein) The three-dimensional folding of the polypeptide chain(s) of a globular protein in its native state.

Third law The entropy or randomness of a perfect crystal is zero at absolute zero.

Transcription The process whereby the genetic information contained in DNA is used to order a complementary sequence of bases in an RNA chain; it involves base-pairing.

Transfer RNA's (*tRNA's*) Structurally and functionally similar species of RNA which have MW near 25,000. Each species of tRNA molecule combines covalently with a specific amino acid; the resulting aminoacyl-tRNA then hydrogen-bonds to a mRNA nucleotide triplet or codon.

Translation The process whereby the genetic information present in an mRNA molecule directs the sequence of amino acids during protein synthesis.

Uncoupling agent A substance (example, 2,4-dinitrophenol) which can uncouple phosphorylation of ADP from electron transport; the energy then is released as heat.

Van der Waals force A weak attractive force, acting over only very short distances, resulting from attraction of induced dipoles.

Viruses Infectious disease-causing agents, smaller than bacteria, which require intact host cells for replication and which contain either DNA or RNA as their genetic component.

Weak bonds Forces between atoms that are less strong than the forces involved in a covalent bond. They include ionic bonds, hydrogen bonds, and Van der Waals forces.

X-ray crystallography The use of diffraction patterns produced by X-ray scattering from crystals to determine the three-dimensional structure of molecules.

INDEX

ABCDE7987654321